Revealing the Heart of the Galaxy

Written in an informal and engaging style, this volume traces the discoveries that led to our understanding of the size and structure of the Milky Way, and the conclusive evidence for a massive black hole at its center. Robert H. Sanders, an astronomer who witnessed many of these developments, describes how we parted the veil of interstellar dust to probe the strange phenomena within. We now know that the most luminous objects in the Universe – quasars and radio galaxies – are powered by massive black holes at their hearts. But how did black holes emerge from being a mathematical peculiarity, a theoretical consequence of Einstein's theory of gravity, to become part of the modern paradigm that explains active galactic nuclei and galaxy evolution in normal galaxies such as the Milky Way? This story, aimed at nonspecialist readers and students and historians of astronomy, will both inform and entertain.

ROBERT H. SANDERS is Professor Emeritus at the Kapteyn Astronomical Institute of the University of Groningen, the Netherlands. Author of *The Dark Matter Problem: A Historical Perspective* (Cambridge, 2010), Sanders has spent his career studying the orbit structure in barred galaxies, active galactic nuclei, and the problem of the mass discrepancy in galaxies. He received his Ph.D. in astrophysics from Princeton University.

Revealing the Heart of the Galaxy

The Milky Way and Its Black Hole

ROBERT H. SANDERS

Kapteyn Astronomical Institute, University of Groningen

CAMBRIDGE
UNIVERSITY PRESS

32 Avenue of the Americas, New York, NY 10013-2473, USA

Cambridge University Press is part of the University of Cambridge.

It furthers the University's mission by disseminating knowledge in the pursuit of
education, learning, and research at the highest international levels of excellence.

www.cambridge.org
Information on this title: www.cambridge.org/9781107039186

First published 2014

Printed in the United States of America

A catalog record for this publication is available from the British Library.

Library of Congress Cataloging in Publication data
Sanders, Robert H., author.
Revealing the heart of the galaxy : the Milky Way and its
black hole / Robert H. Sanders, Kapteyn Astronomical Institute, University of Groningen.
    pages   cm
Includes bibliographical references and index.
ISBN 978-1-107-03918-6 (hardback)
1. Black holes (Astronomy)   2. Milky Way.   3. Galactic center.   I. Title.
QB843.B55S26   2014
523.1′13–dc23   2013024124

ISBN 978-1-107-03918-6 Hardback

Contents

Acknowledgements

In writing this book I have benefited from the efforts of several colleagues who have read and commented on selected chapters. Specifically, I thank Jacob Bekenstein, Harvey Liszt, Mark Morris, and especially Reinhard Genzel.

Over the years I have collaborated or had numerous discussions with colleagues on the subject of the Galactic Center and more generally on active galactic nuclei. Some remain friends and some have vanished – some, sadly, forever. I deeply appreciate these associations and all that I have learned. In particular, I gratefully acknowledge Lyman Spitzer, Kevin Prendergast, Jan Oort, Nick Scoville, Piet van der Kruit, Bruce Balick, Ed Spiegel, Lo Woltjer, Ron Ekers, Tom Bania, Don Backer, David Allen, and Roelof Bottema.

I have made several attempts at observing gas motions toward the Galactic Center with, fortunately, competent collaborators: Gerry Wrixon, Harvey Liszt, and Butler Burton. I am not so sure that I contributed much worthwhile, but it has been fun. I especially enjoyed those nights in Tucson, at the Plaza, after coming down from the millimeter wave telescope on Kitt Peak.

I thank all of those busy scientists, about 30 in all, who have taken time to give their permission for the use of figures and diagrams and, in several cases, provide higher quality images. I am especially grateful to Steve Jurvetson for the inspirational front cover photograph of the Milky Way and to Neil Killeen, Fred Lo, and Miller Goss for the remarkable radio continuum images of Sagittarius A on the rear cover.

I am very grateful to Vince Higgs of Cambridge University Press for his encouragement, support, and many helpful suggestions. He has been the perfect editor.

Finally, I thank my wife Christine for her continued patience with an old guy who just cannot stay away from the office.

1

Introduction: The Luminous Pathway

Nothing can be more spectacular than the nocturnal moonless sky, far from big city lights and haze. This is especially true in the summer, or so it seemed to me in the small Texas town of my childhood. Even in that soupy semitropical climate near the Gulf of Mexico there were nights when the sky was brilliant with stars of all brightness and colors. It appeared to me not as a flat canopy but as a void with depth – three-dimensional – the brighter objects so close that I could almost touch them and the fainter ones fading away to infinity. And through it all ran the luminous band of the Milky Way flowing north to south from Cassiopeia down to Sagittarius with a conspicuous bifurcation halfway between in Cygnus. What was it? What comprised this shining ribbon, present in the same form night after night in the summer sky? I think that it was the appearance of the Milky Way that stimulated my early obsession with astronomy, an obsession that I never outgrew.

Primitive peoples gazed upon this same celestial spectacle and, given their intimate proximity to nature and the absence of interfering human sources of light, were certainly more aware of its appearance and constancy than are we. Although we have no idea of the mystical or superstitious or anthropomorphic associations prehistoric humans assigned to this phenomenon, the mythology of ancient peoples does present several consistent images invoked to explain, or describe, the Milky Way. The first is that of spilt or wasted milk, but milk with a luminous or magical quality. A story out of Greek mythology is that Zeus, having fathered a son, Hercules, with the mortal Alemene wished him to be endowed with god-like qualities, and so, as his wife Hera was sleeping, placed the infant on her breast to suckle the milk of gods. Hera, apparently not an abnormally heavy sleeper, awoke and pushed the strange infant away, spraying the divine liquid across the heavens. For the Egyptians the Milky Way was also a pool of cow milk produced by a heavenly herd evidenced by the stars. And then there is the image of a stream or river. For Hindus it was the Ganges of heaven; for the Australian aboriginals it was the sky river with creature dwellings along its bank. The image I prefer is that of a pathway or heavenly street as described by Ovid in Metamorphosis:

> When the nighttime sky is clear, there can be seen
> a winding highway visible in heaven, named
> the Milky Way, distinguished for its whiteness.
> Gods take this path to the royal apartments
> of Jove the Thunderer; on either side
> are the palaces with folding doors flung wide,
> and filled with guests of their distinguished
> owners; plebeian gods reside in other sections,
> but here in this exclusive neighborhood,
> the most renowned of heaven's occupants
> have *their* own household deities enshrined;
> and if I were permitted to speak freely,
> I would not hesitate to call this enclave
> the Palatine of heaven's ruling class.
> (translation by Charles Martin)

So the Milky Way is the main street of an extremely upscale neighborhood – a silvery Mulholland Drive, meandering its way through a celestial Beverly Hills where the stars dwell.

When I was about nine or ten years old, I read that the appearance of the Milky Way was actually due to uncountable faint stars along the line of sight in the great disk of our Galaxy – stars too faint to be seen individually but, combined, created the appearance of a continuous band of light. But this knowledge remained rather theoretical until age twelve, when my father gave me a small refracting telescope (40-mm aperture). When I pointed the telescope at the Milky Way I actually saw these countless stars filling the field of view. I can only image how startled Galileo must have been when he turned his small self-made telescope to the Milky Way and became the first human to discern its true nature.

We live in a spiral galaxy, a great disk of stars turning ponderously about its center supported against its own gravity by centrifugal force. All the stars that we can see at night belong to this vast stellar system; the stars we can individually discern are close to the Sun and do not appear to be confined to the disk. My early perception of depth in the sky was somewhat of an illusion, because the brighter stars are not generally closer; they are more luminous. The disk contains not only stars but also a very tenuous gas out of which new stars are continuously being formed. Within this gas there are small solid particles, dust that obscures the light of stars behind. It is this interstellar dust in the plane of the Milky Way that creates the appearance of the "great rift" in Cygnus – the apparent splitting of the Milky Way. This dust obscures all visible light beyond a few hundred light years.

Up to the early twentieth century, before astronomers knew about the dust and appreciated its importance in dimming the star light, they perceived that we were near the center of this disk, a somewhat privileged position. Distributed more uniformly around the sky, not lying in the plane of the disk, were various other fuzzy objects such as globular star clusters and "spiral nebulae." But the Milky Way disk with us at the center was perceived to be the principal constituent of the Universe in this rather non-Copernican world-view.

Now that astronomical distance can be determined with precision we understand that the Milky Way is only one of these "spiral nebulae," or rather *galaxies* – actually great star systems themselves lying at previously unimaginable distances from our Galaxy. Moreover, now that the effect of the interstellar dust is understood and quantified, we realize that we are not at the center but at the outer edge of the disk; the center lies 26 000 light years away from us in the constellation of Sagittarius. We can lift this veil; we can peer through the dust by looking at electromagnetic radiation of longer wavelength than visible light – radio waves and infrared radiation – and directly observe the large-scale structure of the Milky Way. As in other distant spiral galaxies we see that the disk of the Milky Way is only one component of the Galaxy. Surrounding the center is a spheroidal system of stars, the "bulge," which is not rotating – at least not so rapidly as the disk; the stars of the bulge are moving with high random velocity.

Proceeding to the center of the Galaxy, the density of stars in the disk and bulge increases. Locally, there is about one star per 30 cubic light years; if this stellar material were smoothed out through space it would amount to about one atom per cubic centimeter – an incredible vacuum. But in the inner parsec of the galaxy, the density of stars has increased to about 300 000 per cubic light year, all moving with random speeds of about 100 km/s. The density of stars is so high here that actual collisions of stars must occur, although not as frequently as we might expect. In the very central region there is one such collision about once per 10 000 years, and the probability of any given star undergoing a collision is small. Even though a collision might be a tremendously violent event, the average rate of producing luminous energy from stellar collision is rather modest: less than 1000 times the luminosity of the Sun or equivalent to a star with ten times the mass of the Sun.

But other even more bizarre phenomena are present in the inner region of the galaxy. There are several thousand massive bright young stars in this region, which implies intense star formation in the past few million years; this is very peculiar given the relatively low density of gas in the inner parsec. And at the very center there is a point-like source of radio emission – emission over a range of microwave frequencies, so-called continuum emission. This source has been designated "Sagittarius A*" or Sgr A* (commonly pronounced "Sag A star"). Although the total power of this radio source is not large, it is very compact – essentially smaller than 100 astronomical units (100 times the distance from Earth to the Sun). Moreover, after correcting for the orbital motion of the Sun about the center of the Galaxy, its position does not change; it remains rock solid at the center. Several of the young bright stars are clearly in elliptical orbits about this radio source just as the planets are in orbit about the Sun. By tracing the orbits over time we can estimate the mass of this object, and it turns out to be more than four million times the mass of the Sun – four million times the mass of the Sun in a region smaller than 100 Astronomical Units (AU) Sgr A* is without doubt the most peculiar object in the Milky Way Galaxy and, indeed, is associated with one of the weirdest constructs conceived by theoretical physics – a massive black hole.

When I finished high school, the only aspects of my life that I was certain of were: (1) I wanted to get away from high school and (2) unlike many of my teenage cohorts, I was not ready to settle down to serious life (job, wife, babies, and so forth). So I looked around and decided to attend the only reasonable and inexpensive university I could find in my neighborhood; that was Rice University in Houston. I remained obsessed with astronomy, but to my father, a solid down-to-earth type, that seemed incredibly fanciful and impractical (little did he realize the long-term pernicious effects of that seemingly innocent gift of a refracting telescope). So I had to hide, or rather camouflage, this interest. I majored in physics, which at least appeared in some respects to be applicable to the real world of jobs and money. At Rice, after flirting with chemistry, mathematics, and philosophy (there was very little else for a heterosexual male to flirt with at Rice in those days) I stayed with physics and, in fact, became quite fascinated with the more theoretical and mathematical aspects (thanks to several motivating teachers such as Stephen Baker). But I remained a closet astronomer. And when it came time for graduate school, I "came out" and made the jump to astronomy at Princeton.

The Princeton University Observatory was a different sociological world than I had known before. It was my introduction to a scientific community that I have previously compared to an extended family. In those days, the senior and dominant figures at the observatory were two remarkable characters: Lyman Spitzer and Martin Schwarzschild. They were both extremely eminent scientists with different but complementary talents. To me, arriving from Texas to the big time world of Ivy League science, Spitzer and Schwarzschild were imposing and, initially at least, a bit scary because to a large extent they had defined modern post–World War II astrophysics. But, at the same time (I quickly discovered), they were very personable and absolutely without arrogance. At Princeton in those days there was much the spirit that we were all marching together sweeping forward the boundaries of knowledge (and graduate students, although foot soldiers, were certainly part of that march).

Astrophysics is, in a real sense, applied physics: the concepts of modern physics are applied to astronomical problems. Moreover, explanations of astronomical phenomena are sought in terms of *known* physics. Occasionally, very rarely, new physics may be discovered, or at least hinted at, by means of astronomical observations. But the first impulse is to seek an explanation in the context of textbook physics. This is probably a good thing because otherwise quite fanciful and unnecessary excursions arise while constructing "theories" underlying observations having a completely conventional explanation. The point is that there is a built-in conservatism to astrophysics; new physics, when it comes, most often comes from the physics community.

So it was at Princeton. The staff at the observatory set very high standards for astrophysics but were inherently conservative with respect to new physical constructs. Across the street, however, at the Palmer Physics Lab, it was a different story. There was a group of wild insurgents led by the chief radical, John Wheeler. Now, John Wheeler seemed nothing at all like a extremist; he was soft-spoken, extremely polite, and, in class, treated all students with great respect, even when they didn't deserve it. But in the realm of ideas he was a true revolutionary. He worked on and coined the expression "black hole."

At that time – 1966, 1967 – the idea of black holes seemed an extremely bizarre concept. Only several groups in the world were thinking about such hypothetical objects: at Caltech, Cambridge and Oxford, Moscow, and Princeton. Now, in the modern world, everyone who is reasonably sentient, has heard the term "black hole" and has a rough idea of what it is: a mass that is so concentrated that nothing, not even light, can escape from it. But actually these objects have even more peculiar properties. In this period of the late 1960s several significant theorems on the nature of black holes were being developed, the most important of which was the singularity theorem of Roger Penrose at Oxford. He proved that, in the context of General Relativity, mathematical singularities may exist in space–time and, in a sense, are inevitable. To physicists this seemed a particularly strange concept. A singularity is a well understood concept where, at a point in space, a function, a mathematical object, "blows up" (approaches infinity) or becomes undefined. That's fine in mathematics, where the function is a pure abstract entity, but that singularities should actually exist in nature, where the function is a real physical quantity such as force or density, is very odd indeed. Penrose had proven that if General Relativity is the correct theory of gravity, and there were and are no experimental contradictions to this theory, then such bizarre points must exist (of course, General Relativity may in some sense be "correct," but it is certainly an incomplete reflection of deeper theory).

Apart from these theoretical developments there had been for three decades the distinct possibility that these objects might actually exist in an astrophysical context. In 1930 Subramanyan Chandrasekhar, a young post-doctoral fellow at Cambridge, wrote his famous paper on the upper limit to the mass of white dwarf stars. White dwarfs are very faint small dense stars of about one solar mass (but only 1/100 of the solar radius) and are certainly the end stage of stellar evolution (the example of one teaspoon of white dwarf matter weighing one ton is almost a truism). Stars are stable objects in which two forces are in balance: the gravity force that pulls all matter of the star toward its center and a pressure force that pushes the other way and keeps the star from collapsing. In a normal star such as the Sun this outward force is provided by the usual thermal pressure of hot gas. But a star that has totally exhausted its nuclear fuel and has no way producing heat energy in its interior may also be supported entirely by the degenerate pressure of electrons, a pressure that exists because the electrons are packed so densely within the star. Chandrasekhar had discovered that it was not possible for this packing pressure to support a star if its mass were greater than about 1.4 times the mass of the Sun; there is no force available to support a more massive star against the force of gravity. A massive star at the end of its life that has exhausted its internal fuel source should therefore collapse to the point where light cannot escape; it should become a black hole.

Most stars in the Galaxy are less massive than about two solar masses, but there are a number of quite massive stars – going up to five, ten, or even 100 solar masses. Some of these massive stars certainly conclude their stellar existence in a spectacular supernova explosion. Supernovae are dramatic events in which a star suddenly brightens by many orders of magnitude (it may even outshine an entire galaxy) and then fades from view over an interval of a month or so. The entire star may be disrupted by this event so no remnant is left. But supernovae do not occur frequently enough; in the Galaxy there are perhaps one or two per century, insufficient to take care of all stars above the Chandrasekhar limit. What happens to all of these these massive stars when the nuclear fuel is gone? To avoid the black hole trap they must either lose mass or somehow get around the Chandrasekhar limit.

When I appeared at the Princeton Observatory in 1966, both of these possibilities were being considered. By looking at ultraviolet spectra of hot young stars, Don Morton had discovered that these stars were actively losing mass; perhaps they could lose enough mass during the course of their active lifetimes to push them below the Chandrasekhar limit before the fuel was exhausted. Jerry Ostriker, who had been a student of Chandrasekhar, was considering the effect of rotation on the mass limit; all stars rotate to some degree and perhaps this rotation could provide some additional support against gravity. This was, in general, the approach of astrophysicists: rather than immediately consider bizarre consequences of gravitational collapse, look for a conventional way out. There was the feeling that Nature, somehow, would avoid the bizarre "unphysical" black hole trap.

As it turned out, these escape exits were blind alleys; early mass loss is insufficient to significantly reduce the mass of young stars and rotation can only marginally extend the Chandrasekhar limit. Moreover, the subsequent discovery of neutron stars – compact objects supported by the degenerate pressure of neutrons rather than electrons – does nothing to alleviate the problem of gravitational collapse; neutron stars also have a mass limit that is comparable to that of white dwarfs. Finally, the development and application of X-ray astronomy in the 1970s produced convincing evidence that stellar mass black holes actually do exist; the evidence was in the form of X-ray emission from massive, otherwise dark objects – more massive than the white dwarf or neutron star limits – accreting gas from a nearby binary companion.

But here I am not going to bother with these puny stellar mass black holes, interesting though they are in establishing the existence of this class of objects. I am going to discuss massive, or supermassive, black holes that almost certainly exist in the dense nuclei of galaxies such as the Milky Way – objects of more than one million solar masses.

Back in 1963 when I was an undergraduate at Rice, Maarten Schmidt, a young Dutch astronomer working at Caltech, succeeded in identifying the spectral lines in a puzzling class of objects that had been discovered several years before. These objects were star-like in their appearance but conspicuous sources of radio emission, subsequently designated "quasi-stellar radio sources" or "quasars" (later it was found that most such objects have low or undetectable radio emission; they are radio quiet). Quasars are not preferentially in the plane of the Milky Way, but are distributed more or less uniformly across the sky, like the distant galaxies (an early hint that they are also extragalactic). The spectra had been measured: there were broad, conspicuous emission lines imposed on a bright continuum, but these lines had not been identified with any known substance. Schmidt's breakthrough was his realization that these lines are actually the common lines arising from ionized hydrogen (and nitrogen) but very highly redshifted – implying a recession velocity of 20% to 40% that of the speed of light. If these objects follow the usual Hubble law (distance proportional to redshift), this means that they are incredibly distant – hundreds of millions of light years – at the time, the most distant objects ever detected. Then, because we can see them at all, it also means that they are incredibly luminous – far more luminous than any source ever discovered, 100 or

1000 times more luminous than all of the stars in a major galaxy like the Milky Way. Yet, they did not appear to be galaxies; they seemed to be star-like. So all of this electromagnetic energy arises in an extremely small region by astronomical standards, perhaps smaller than 1 light year. When it was discovered that the total flux of radiation from quasars varied on timescales of weeks or months, then, because they could not be larger than the distance traveled by light on this timescale, the size was even more drastically constrained to less than a light year.

At the time this appeared to stretch the known laws of physics. How could so much energy be emitted from such a compact region? What was the source of all of this electromagnetic radiation ranging, as we now know, from radio emission to gamma rays? Most astrophysicists at that time did what astrophysicists do: they tried to find an explanation in terms of conventional physics (not all did; some proposed that the phenomenon was due to totally new physics of unknown basis – matter recently created and shot out of galaxies, matter with an intrinsic redshift beyond that due to the Hubble expansion).

Lyman Spitzer had developed a "conventional" model. It was generally known that as an isolated stellar system (a star cluster or a galactic nucleus) evolves, it loses stars. Because the ejected stars carry away energy, the remaining system contracts and becomes denser; in time its density approaches infinity. Spitzer realized that before this happened, the stars in the system would actually begin to collide and this could possibly liberate a large amount of energy.

I was the third Spitzer student who worked on this problem (following Bill Saslaw and Mike Stone). Spitzer's idea was that the collisions would, by and large, disrupt stars; the gas so liberated would cool by normal thermal radiation and collapse to the center of the system. Because of residual angular momentum, the gas would settle into a compact disk, new stars would form and diffuse out of the disk, and the entire system would contract further, accelerating the whole process (the end result of this process was not discussed). The hope was that quasar luminosities could be produced by such a mechanism. But a problem with Spitzer's model was that the radiation produced was thermal emission from hot gas, whereas the continuum emission observed from quasars was nonthermal (due to relativistic electrons spiraling in a magnetic field). The radiation predicted did not match the radiation observed, and so the model required further "processing" of the radiation outside the dense stellar system.

Then, in 1967, Stirling Colgate, a nuclear physicist, pointed out that before stellar collisions became frequent and disruptive, they would become frequent and soft; these soft collisions would lead to coalescence, not disruption, of the colliding stars and more massive stars would be built up. Colgate estimated that this coalescence process would cease when the stars became more massive than about fifty solar masses (this would happen because the stars are bloated by the extra collision energy), and these fifty solar mass stars would then explode as supernovae on short order. The greatly enhanced supernovae rate would produce the high luminosity and all of the nonthermal emission observed from quasars. Colgate's model appeared to be quite plausible because it addressed the actual observations of quasars.

For my Ph.D. dissertation work, Spitzer asked me to look into this question: does a system of colliding stars lead generally to disruption of the stars and liberation of gas, or to coalescence with increasing average stellar mass? I found that the answer depends somewhat on the properties of the system; for a certain reasonable range of properties coalescence would indeed dominate, but I concluded, contrary to Colgate, that the growth would not stop at 50 solar masses; it would accelerate and not be terminated by usual supernovae explosions – a sort of runaway growth of stellar mass leading finally to one or several very massive stars. I did not speculate what would happen to these massive stars, but they are known to be highly unstable to gravitational collapse.

As later summarized by Martin Rees (Cambridge), this is only one of several possible paths leading to the formation of a massive black hole in a galactic nucleus. It would seem that the formation of black holes, and very big ones, is a natural development in dense galactic nuclei.

But black holes are, after all, black. They are not supposed to shine. How can they be the source of all of this electromagnetic radiation from quasars? The answer is that we do not, of course, see the black hole itself; we see matter being accreted onto a black hole. Before the accreted mass disappears from view, it can radiate; in fact, it may emit up to 10% or 20% of its rest mass as electromagnetic radiation. But luminosities of 100 times that of an entire galaxy would still require something like the mass of ten suns disappearing down the hole every year. So not only is the black hole necessary for the quasar phenomenon, but also a copious supply of "fuel" must be available in the near environs of the hole. Moreover, if quasars live for ten million years (a reasonable supposition given the size of the extended radio structure surrounding quasars) then the black hole must have a mass of one hundred million solar masses.

In 1969, Donald Lynden-Bell (Cambridge), in a seminal paper published in the journal *Nature*, stressed the observational fact that the co-moving number density of quasars (i.e., after taking into account the expansion of the Universe) seemed to be higher in the past than in the present (more and more quasars are seen at large redshift or far in the past). If the quasar phenomenon is due to accretion onto massive black holes in galactic nuclei, then where are all of the dark black holes now? Because massive black holes do not disappear or lose mass, the nuclei of normal galaxies should also contain black holes. Because they are not shining, they must be underfed; these objects are suffering from a fuel crisis.

This, at the time, was quite a radical suggestion. It was known that 2% or 3% of nearby galaxy nuclei are indeed brilliant and are likely to host massive black holes. In 1943 the American astronomer Carl Seyfert had discovered and catalogued several spiral galaxies with bright star-like nuclei and broad emission lines coming from a compact central region. When quasars were later discovered, these Seyfert galaxies seemed to belong to this same class but of lower power. But Lynden-Bell was proposing that most galaxies have gone through Seyfert or quasar episodes. Perhaps black holes are a *normal* component of galactic nuclei and emerge naturally during the course of galaxy formation or evolution. Perhaps the Milky Way contains one of the bizarre objects at its center.

After Princeton, I went on to my first actual job as a post-doctoral fellow at Columbia University in New York City. The staff at Columbia was an eclectic collection of very bright but somewhat cynical characters who created quite a different atmosphere than at Princeton (at the time, I had tired of the generally wholesome boy scout attitude at Princeton and was quite ready for this dose of healthy cynicism). The caustic tone of Columbia was probably related to the surrounding atmosphere of New York, where one continually faced potential assaults on body and soul – but it was exciting. My immediate boss was Kevin Prendergast, a native New Yorker and a charter member of the cynics club, who also happened to be a brilliant astrophysicist and applied mathematician. He had developed a very clever technique for simulating gas dynamics on a computer, the "beam scheme," and he asked me to adapt and apply this method to a specific problem: the focusing of explosions in thin gaseous disks. At the time, explosive phenomena in galaxies – apparently evidenced by so-called "radio galaxies" (in which the optical galaxy was found between two large lobes of radio emission), the more nearby example M82 (supposedly an "exploding galaxy"), and large noncircular gas motions observed in the Milky Way in the direction of the Galactic Center – were quite in fashion (it was an explosive time politically as well). The immediate question was: Could a violent explosion in the center of a thin disk be narrowly focused in two opposite directions to explain the morphology of radio sources? Or could such an event excite noncircular gas motions several thousand light years from the Galactic Center? As it turned out, these speculations were something of a detour in understanding the dynamics of both normal and active galaxies, but this was not apparent at the time. The main road led directly to black holes and their off-and-on fueling.

If Lynden-Bell's suggestion that normal galaxies contain massive black holes were true then even our Galaxy, the Milky Way, will contain a massive underfed black hole and efforts should be made to find it. But if it doesn't shine, how can we find the black hole in the center of the Milky Way? Lynden-Bell suggested that we should look for dynamical evidence for a mass concentration at the Galactic

Center, perhaps from rotation curves – the dependence of rotation velocity on distance from the center. We should also expect to find nonthermal radio emission connected with this such an object. In fact, as it turned out, both signatures have been identified in the Galactic Center.

After 30 years of research, the big black hole in the Milky Way has unquestionably been found, and the remarkable story of this search and discovery is the story that I will tell here. The reality differs from Lynden-Bell's original vision, but that is not surprising for innovative and speculative ideas. The object actually has a mass of four and a half million solar masses, considerably less than 100 million solar masses proposed by Lynden-Bell; so the Galaxy could never have been a quasar. The Center of the Galaxy has probably been and will be again an "Active Galactic Nucleus" – a conspicuous and powerful source of electromagnetic radiation. This is because the fuel for activity, the accretion of gas into the deep well of the black hole, is not smooth and continuous. There seem to be separate random accretion events when the black hole flares briefly. It is like a sleeping but still active volcano. Mostly the hole is quiet, but it will occasionally erupt and spew fire.

The actual story is larger than this one discovery. In 1970, the idea that black holes exist – let alone *massive* black holes in the centers of galaxies (power plants for the most extreme sustained sources of radiation in the Universe) – was highly speculative and taken seriously only by a few precocious individuals. But by 1980, most astrophysicists thought that this was probably true; by 1990, it had become the paradigm. It is another one of those revolutionary developments in science where, in a relatively short period of time, a speculative and bizarre model becomes the conventional dogma – a major paradigm shift. In that sense it is similar to the issue of dark matter: over an interval of several years the entire community of astronomers went from believing that the Universe is essentially visible to our eyes and instruments to being convinced that most of the material world consists of hypothetical undiscovered particles detectable only through their gravitational influence.

In the case of black holes, a major contribution to this change in perception – how black holes went from being a highly theoretical construct and toy for physicists like John Wheeler to actual astrophysical objects with conspicuous observational consequences – has been the discovery of the black hole in the center of the Milky Way. Here I will describe this development from a historical perspective. I include a general discussion of the very peculiar properties of black holes, possible scenarios for the formation of black holes in galactic nuclei, and a description of the observations leading directly to the discovery of the Milky Way black hole. Again the style is narrative and based to a large extent on my personal experience, so it is necessarily somewhat biased. Because scientific progress is not a straight line, I will mention several detours along the path to the Milky Way black hole and discuss the significant contributions of a few individuals who are not usually mentioned in this story.

My personal history has led me from New York, by way of Charlottesville, Pittsburgh, and Geneva, to the Kapteyn Institute in Groningen, a provincial town in the north of the Netherlands. The institute is named for the scientist who, a century ago, began the systematic study of the Milky Way using methods of classical astronomy: careful measurement of the positions and motions of stars. Although Jacobus Kapteyn came to wrong conclusions about the scale and position of the Sun in this giant star system, his data were used by his famous student, Jan Oort, to determine the true extent of the Galaxy, to describe its rotational motion, and to estimate it mass. Oort discovered the direction and distance to the Center. He was able to identify the Center of rotation with a strong radio source in the constellation of Sagittarius, and he found, using the new technique of radio astronomy, gas features with peculiar velocities in the direction of the Center, features that he thought (incorrectly as it turned out) were indications of vast explosive events in the Center. But his work inspired younger followers and led to many of the major discoveries described here. In a real sense Oort discovered the Center of the Galaxy.

So parallel to describing the discovery of the Milky Way black hole, I consider first the discovery of the Milky Way as a galaxy. After all, this luminous band of silvery light in the summer sky is the path that leads directly to the black hole. And more than that, it is for me and for others, a source of youthful fascination with astronomy and an inspiration for peering deeper and further beyond its obscuring veil.

2

The Discovery of the Milky Way Galaxy

2.1   Kapteyn's Universe

The "hondsrug" or "dog's back" passes for a mountain range in the Netherlands. It is a ridge of sand reaching an altitude of 30 m and stretching southeast to northwest from the German border through the wooded province of Drente into province of Groningen. In fact, the most northern point is the city of Groningen, which is certainly the reason why a city is there: it is the closest point to the North Sea that is still above sea level – that is to say, on a natural geological formation. To the north of the city there are ancient small villages built on "terps," artificial small hills created over centuries from animal and human waste, usually with a church at the highest point. Long before the construction of dikes, local farmers would gather on these terps during storms or exceptionally high tides. It is a wet and grim climate: in winter low clouds hang over the flat green treeless landscape; sunny summer days can be disrupted by sudden downpours – soaking cyclists and sending them scurrying for bridges and highway overpasses.

Groningen is a large provincial town – the central market city of this rural region. It is, by Dutch standards, rather isolated – 200 km from Amsterdam and the other metropolises of crowded Holland. Basically it bears the same relation to the Netherlands as does Novo Sibersk to Russia; from the point of view of Holland, Groningen is in the far frozen and gloomy North. But because of this relative isolation, it has developed its own dialect and culture and bustling student life. For as unlikely as it might seem, the city has a university.

The University of Groningen was established in 1614 by the city and provincial administration, making it, after Leiden, the second oldest university in the Netherlands. From an American perspective it is quite venerable, predating Harvard by twenty-two years, but compared to the medieval universities of Europe – Bologna, Oxford, Paris, Cambridge – it is a relative newcomer. The original class consisted of about 100 students specializing in the staple topics of universities of the time: divinity, law, medicine, philosophy. By the middle of the nineteenth century the student body had no more than doubled in size and there was talk of closing the university. But in 1876 the Dutch government began a modernization of Dutch universities; as part of this process, chairs of astronomy were established at each of the three, by then, state-subsidized universities – Leiden, Utrecht, and Groningen. So Groningen got its professor of astronomy, Jacobus Kapteyn, who at the age of twenty-seven gave his inaugural lecture and began his function as the astronomer of Groningen.

Jacobus Cornelius Kapteyn (Figure 2.1), "Ko" for short, grew up in a large family (fifteen children) in the village of Barneveld near Utrecht, in the center of the country, where his father kept a boarding school for about fifty pupils. He and his siblings were imbued with traditional Calvinist values of thrift, hard work, discipline, and humility – virtues he would need in Groningen because the university, after appointing him professor, refused to give him the observatory that had been promised. In those days, an astronomy institute without an observatory was unheard of; to use a modern simile, it would be like a pizzeria without an oven – it was just not possible to do the job, or so it seemed. Kapteyn could have easily despaired, given up, looked for another job – but that was not in his character. If he did not have the necessary observatory he would use other observatories in more favorable climates. He would make arrangements with foreign astronomers to use their observational facilities and accumulated data, and in return he would offer those resources that he possessed: competence and hard work. His observatory would become an astronomical laboratory for the analysis and interpretation of data (see the excellent discussion of the life and career of Kapteyn in Groningen by Wessel Krul).

Science is a social activity and the various scientific sub-fields form micro-societies. Because of his relative isolation in Groningen, Kapteyn was forced, possibly more so than had he been in a more vital center of activity, to reach out to the larger international fraternity of astronomers. His first arrangement was with David Gill at the Royal Observatory at the Cape of Good Hope. Kapteyn measured the positions of stars in the Southern Hemisphere on photographic plates taken by Gill – in fact more than 450 000 stars – and it took him twelve years. The resulting catalogue, the "Cape Photographic Durchmusterung," brought Kapteyn to the attention of the great American observatory builder George Ellery Hale, and thus began, after 1908, Kapteyn's collaboration with American astronomers at the newly constructed Mt. Wilson Observatory, where, on annual visits, he continued his program of measuring the position and motion of stars.

Now what was the point of all of this detailed and tedious work? Was Jacobus only a dull plodding measurer of plates? In fact, he did have a grander idea.

Figure 2.1 An older and thoughtful Jacobus Kapteyn. (Courtesy of University of Groningen.)

His purpose was to determine the structure and motion of the universe of stars, that is, the Milky Way, which was thought by most astronomers at that time to comprise the Universe. It was known that in the context of Newtonian gravity there were problems with an infinite homogeneous Universe (realized by Newton himself). The gravitational potential would be indeterminate and the gravitational force would be infinite. One could say that with homogeneity the infinite gravitational force on an object from all directions would cancel, but then this would require perfect homogeneity – any slight deviation would lead to rapid collapse. Two solutions presented themselves: one was to modify Newtonian gravity by adding an exponential cutoff that became effective at large distances (this explanation was considered by Hugo von Seeliger, Kapteyn's competitor for the professorship at Groningen), and the second, less radical, was to propose that the Universe was not, in fact, homogeneous. This was the direction pursued by Kapteyn: to solve the Sidereal Problem.

Kapteyn was not the first astronomer to try to deduce the structure of the system of stars containing the Sun. It was known since Galileo first turned his telescope to this luminous band in the sky that the Milky Way was a great concentration of stars along the line of sight. Thomas Wright, the English surveyor and natural philosopher, proposed in 1750 that the Universe consisted of an enormous spherical shell of stars, like the surface of a balloon. The Sun was one of these stars on the shell and from this vantage point, looking tangentially along the shell, one sees many more stars than looking radially outward or inward – thus, we observe the Milky Way as a band of stars across the sky. The position of the Sun on the shell was irrelevant; there was no center or edge.

This was taken up by the German philosopher Immanuel Kant in 1755, who reportedly read an inaccurate review of Wright's book and interpreted the proposed structure as a disk rather than a spherical shell (a very fortunate misinterpretation). Kant proposed that this system of stars was flattened and supported against its own gravity by rotation – that the disk of stars had formed from a larger gas cloud that collapsed and flattened as it rotated faster. Kant went on to speculate that the other "nebulae" distributed more uniformly around the sky could be very distant versions of the Milky Way; viewed at random angles they appeared flat (edge-on) or rounder (more face on) – altogether an amazingly modern view but in conflict with the idea of the star system, the Milky Way, comprising the entire Universe. Kant's model of the structure of the star system was placed on an observational basis in 1785 by the great English (German born) astronomer William Herschel, who, using methods of star counts later emulated by Kapteyn, deduced that the Sun was somewhere near the center of the disk because the density of stars in the plane of the Milky Way seemed to fall off uniformly in all directions.

Kapteyn's extensive analysis of a much larger body of data and a more sophisticated statistical analysis made this result more quantitative and precise. The conclusions drawn from his enormous project of plate measuring and analysis were published after more than forty years in 1922, the year of his death: "First attempt at a theory of the arrangement and motion of the sidereal system." The Milky Way disk was about 40 000 light years in diameter, with the Sun being not precisely at but relatively close (about 2000 light years) to the center. Kapteyn had also studied the motion of the stars in this system: in the neighborhood of the Sun there appeared to be two streams of stars moving in opposite directions with velocities of about 50 km/s. This was consistent with centrifugal support of the system against its own gravity. The whole picture became known as the Kapteyn Universe and it was a magnificent construct – magnificent but wrong.

2.2   Pickering's Harem

In the 1890s the director of Harvard College Observatory, Edward Pickering, hired a number of women (human computers) to perform the necessary routine work of an observatory – the sort of work that Kapteyn had been carrying out in Groningen. In those days women were not allowed to observe with

Figure 2.2 Henrietta Leavitt, the discoverer of the period–luminosity relation for Cepheid variable stars, at work in Harvard College Observatory. (AIP Emilio Segre Visual Archives, Physics Today Collection.)

telescopes – that was clearly work for men – but it was thought that they were very suitable for tedious routine work such as measuring the relative brightness and positions of stars on photographic plates – work that was too dull to hold the interests of serious male astronomers.

In fact, there were a number of highly talented and well-trained women in this group, any one of whom could have become a major astronomer at a major observatory had the attitudes of the time permitted. One of these female "computers" was Henrietta Leavitt (Figure 2.2), a graduate of Radcliff College, who from the time of her university education, had an interest in astronomy. Her remarkable personal story has been very well told elsewhere (see "*The Big Bang*" by Simon Singh), and here I only describe her singular achievement. She compared photographic plates of the Large Magellanic Cloud (LMC) taken over intervals of time at Harvard's southern station in Peru. The LMC, now known to be a dwarf companion galaxy to the Milky Way, contains many bright stars that can be discerned individually. Leavitt found a number of periodic variable stars (stars that regularly vary in brightness), called Cepheid variables after their nearer prototype Delta Cephei, and was able to determine their period of variation, usually between one and sixty days. When she plotted the average apparent brightness of these stars against the period she found a well-defined relationship – the longer the period, the brighter the star (Figure 2.3). By assuming that all of the stars in this system were about at the same distance (that the LMC is small compared to its distance) she deduced the existence of the period-luminosity relationship, the luminosity

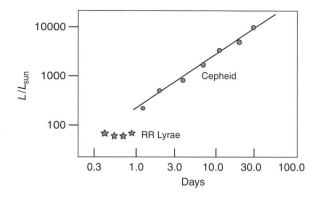

Figure 2.3 A schematic view of the period-luminosity relation for Cepheid and RR Lyrae variables. The luminosity (*y*-axis) is in units of solar luminosities.

being the true power of radiant energy emitted by the star. This relationship was so precise that it could be used as an astronomical yardstick, a distance indicator: simply measure the period of a Cepheid, deduce its luminosity or true brightness, compare that with the apparent brightness, and you have its distance. All that it is needed is the calibration of the relation by precisely measuring the distance to one Cepheid.

The significance of this discovery cannot be overstated. It is arguably the most important astronomical discovery of the twentieth century because it led directly to the work of Shapley and Hubble in establishing the scale of the Milky Way and its relation to the Universe and indeed the scale of the entire Universe. The far-reaching significance was very quickly realized, as is evidenced by the fact that Leavitt was allowed to publish this result under her own name in the *Annals of the Harvard Observatory*. It also shifted the center of astronomical research on the Milky Way and on cosmology from Europe, specifically from Kapteyn's astronomical laboratory in Groningen, to the United States.

2.3   Shapley's Supergalaxy

At the time of Leavitt's discovery, Harlow Shapley was a student of the great American astrophysicist Henry Norris Russell at Princeton University. He had worked on another sort of variable star – eclipsing binaries in which a fainter star, in orbit about a brighter star, periodically eclipses and is eclipsed by the brighter star. Shapley realized, along with Russell (and by Leavitt before them), that Cepheids could not be of this sort: the variations must be intrinsic; a Cepheid variable must be a single pulsating star that is an accurate clock with its period related to its internal properties as is evidenced by the luminosity. He realized that he could apply Leavitt's period-luminosity relation to measure the size of the Sidereal system – the great stellar system of the Milky Way. He only needed a proper observatory to do the work.

The enthusiastic, ambitious young Shapley arrived at Hale's Mt. Wilson observatory in the spring of 1914, where he began a large project mapping the three-dimensional distribution of globular clusters, making use of the period-luminosity relationship for Cepheid variables. Globular star clusters are gravitationally bound systems of roughly 100 000 stars and are distributed uniformly around the sky, that is, they are generally not found in the plane of the Milky Way. It had been known for some time that these systems contained variable stars – intrinsic variables – of rather shorter period (less than one day) than the classical Cepheids studied by Leavitt in the LMC. Shapley assumed that these cluster variables were on the same period-luminosity function as the Cepheids and began deriving the distance to the various globular clusters. Leavitt's observation of the period-luminosity relationship for Cepheids in the LMC could give not absolute distances but relative distances; the relationship was uncalibrated because no one knew the distance to the LMC. Shapley's first task was to calibrate the period-luminosity relation by independently determining the distance to several Cepheids and thereby their luminosity.

The first attempt at calibration was by the Danish astronomer Ejnar Hertzsprung (Kapteyn's son-in-law). He used a method known as statistical parallax (using the average motion of the Sun with respect to nearby stars to provide a baseline) to estimate the distance to several relatively nearby Cephied variables in the disk of the Milky Way. His estimated distance scale was far off the mark; it was much too small. Shapley corrected this estimate (for example, he removed two Cepheids from Hertzsprung's sample because their light curves were atypical), and came out with his more precise calibration. On this basis, and making the assumption that the period-luminosity relation for the globular cluster variables was the same as that of the Cepheids, he was able to map the three-dimensional distribution of the globular clusters, and his results were presented in 1918. He found that these objects were distributed in a great spheroidal system around the plane of the Milky Way and that the center of this system was 60 000 light years away from the Sun in the direction the constellation Sagittarius; in other words, the Sun and its system of planets was far from the center of this great configuration of stars delineated by the globular clusters – far from the center of the Universe. Moreover, Shapley estimated that entire Galaxy is vastly larger than had previously been imagined, about 300 000 light years in diameter. This, after Copernicus, was another blow to the self-centered view of humans.

Kapteyn had met Shapley at Mt. Wilson during his visit of 1914 and was impressed with this brilliant and energetic young astronomer (Figure 2.4). But

Figure 2.4 Harlow Shapley as he must have looked at the time he met Kapteyn. (*Ad Astra Per Aspera, Through Rugged Ways to the Stars*, by Harlow Shapley. New York: Charles Scribner's Sons, 1969. Courtesy AIP Emilio Segre Visual Archives, Shapley Collection.)

naturally he was unhappy with Shapley's arrangement of the stellar universe, particularly its large dimension and the very peripheral position of the Sun. He found this inconsistent with the observation of the observed symmetry of star counts in all directions. Kapteyn specifically argued that the use of only eleven Cepheids with imprecise distance estimates to calibrate the period-luminosity relation was dangerous and that the assumption that the cluster variables (now known as RR Lyrae stars) were on the same period-luminosity relation as the non-cluster variables (the classical Cepheids) was unjustified. He was correct in both of these criticisms, but, in fact, these two effects largely canceled one another: the error made in calibrating the Cepheid distance scale was offset by that of placing the cluster variables on the same period-luminosity relation as the classical Cepheids.

Shapley, in a sense, believed in the Kapteyn Universe but thought that it was a "local" system and part of the larger supergalaxy (Figure 2.5). How else could he accommodate his assumption of no absorption with the uniform decline in stellar density in all directions from the Sun? Absorption of starlight was the critical issue. It was known that there was obscuring matter in space between the stars; there were dark patches in the Milky Way, areas where no stars were seen (such as

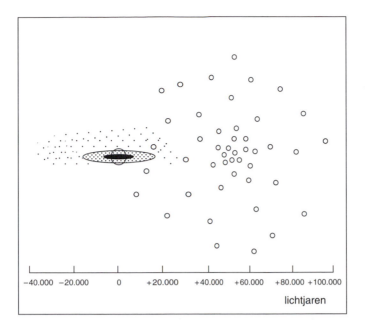

Figure 2.5 The Kapteyn Universe and its relation to the Shapley supergalaxy as mapped by the system of globular clusters. (From *Kosmos* by W. de Sitter, 1934.)

Barnard's cloud), but it was not clear if this absorbing material, presumably comprising small particles, was distributed generally throughout the space between the stars.

Kapteyn was aware of and worried about this possibility and even estimated the mean absorption along the line of sight in the Milky Way. But Shapley, in fact, presented evidence against the importance of absorption. It was known that light is not only absorbed by small particles, but it is also reddened; that the blue light is preferentially scattered away and only the red light proceeds relatively uninhibited from the star to the observer. Shapley had failed to find evidence for this reddening in his globular cluster observations; more distant clusters were no redder than closer clusters. This certainly must have been a relief to Kapteyn. It became apparent only later that absorption was significant but concentrated toward the plane of the Milky Way; it played less of a role for objects such as globular clusters lying outside of the Milky Way disk.

Like Kapteyn, Shapley thought that the Milky Way, the system of stars, was the Universe, and that the other fuzzy objects outside of the plane of the Milky Way, the "spiral nebulae," were satellites like the globular clusters and not really very distant compared to the size of the Milky Way itself. This became the topic of the Great Debate in 1920 between Shapley and Heber Curtis, director of the Lick Observatory in California. What actually was the Universe? Was it the Milky Way and its nearby satellites, as argued by Shapley, or was it, following Curtis, vastly larger with the spiral nebulae being very distant and enormous galaxies like the Milky Way?

The arguments given in the Great Debate go beyond the aim and scope of the discussion, but the issue was a fundamental clash of world views; at stake was nothing less than the human view of the scale and structure of the Universe. It was the seminal event and perfect example of the dialectic process by which science progresses. Indeed, much of human endeavor advances by just such Hegelian conflict between ideas, but science offers the means of choosing between alternatives – direct observation and experiment. The road, however, is never straight and a detour was provided by Adriaan van Maanen (who spent a short time in Groningen in 1911 before going on to the United States and finally Mt. Wilson Observatory). van Maanen claimed to have seen the Andromeda galaxy spinning through the proper motion of its stars; if it were really a distant galaxy as large as the Milky Way, then it would have been spinning faster than the speed of light. On the other side there was the young astronomer Edwin Hubble, who in 1922–1923 observed Cepheid variables in the Andromeda Galaxy and provided strong evidence that this object was more than 1 000 000 light years away – a previously unimaginable distance. Normally, the direct geometrical evidence provided by van Maanen's observations would be preferred to a poorly understood empirical relation between period and luminosity of a class of variable stars, but, as it turned out, van Maanen's measurements were beset by systematic effects and discredited.

So Hubble, and Curtis before him, were right – that, in fact, the Universe is vastly larger than the Milky Way; that Immanuel Kant had been correct in supposing that the spiral nebulae were island universes; and that Shapley and Kapteyn were certainly wrong in their view of the stellar Universe comprising only the Milky Way. The Milky Way is a spiral galaxy, one of very many, rotating about its center with a speed in excess of 200 km/s. Kapteyn had indeed underestimated the effect of interstellar absorption, and the Sun was certainly in an obscure peripheral position some 26 000 light years from the true center (correcting Shapley's results for his neglect of absorption shrunk his Milky Way by a factor of three). Kapteyn's two streams moving in opposite directions resulted from the fact that the stars near the Sun are on elliptical orbits, with some moving toward the center and some moving away with respect to the frame moving with the Sun about the center of the Galaxy.

This implies that the Universe of galaxies can on a large scale be homogeneous; the galaxies can be smoothly distributed in space. Then what about the problem of Newtonian cosmology in a homogeneous Universe – the fact that the gravitational potential is indeterminate and the force is infinite? By 1920, this problem had also vanished with the publication five years earlier of the *General Theory of Relativity* by Albert Einstein. Einstein's theory of gravity permitted a class of homogeneous cosmologies (but did not permit a stable static cosmology). The Universe could indeed be a infinitely large and smooth and filled with galaxies such as the Milky Way as long as it was dynamic, that is, expanding or contracting.

2.4   The Legacy of J. C. Kapteyn

Kapteyn did not live to see his model Universe so thoroughly discredited. After the First World War he became an early opponent of the punitive exclusion of German scientists from international organizations. He felt that science is a truly international endeavor and should transcend international politics, and this attitude resulted in his increasing isolation from his colleagues in Britain and the United States – after 1914 he never again visited Mt. Wilson. He retired in 1922 at age seventy, after the Great Debate but before the issue was resolved. In the last few years of his active professorship he and his wife had moved from their country home in Drente to a hotel on the Grote Markt, the central marketplace of Groningen, and after his retirement left quite suddenly without making the traditional farewell lecture given by professors at that time. Perhaps he never really felt at home in this remote and somewhat obscure provincial town at the northern end of the hondsrug. He never returned and died in Amsterdam within a year, a man saddened through rejection – although not for scientific reasons – by the international community of astronomers. This rejection must have been indeed painful for a man whose career was built upon just such connections with the larger family of astronomers.

Kapteyn was the ultimate empiricist. As reflected in his life work, he believed that the way to truth was through careful observations and measurement. Now he is often thought of as being misled and misleading the community about the nature of the Milky Way and the Universe at large. But he was more sophisticated; he was very aware that the possibility of obscuration by dust particles could be hiding the true extent and shape of the Galaxies and of the position of the Solar System within it. And as later work would bear out, his precise measurements of stellar positions and motions would permit an accurate assessment of the position and distance to the Galactic Center and the rotation of the disk about that center.

In a more general sense, he had a lasting influence on Dutch and thereby world astronomy. More than anyone, he established the international character of astronomy in the Netherlands, which is very important for a small country with limited resources; this internationalization has led to the phenomenon that young Dutch astronomers are a leading export after cheese and tulips. He developed the style of interpretive astronomy, between observations and theory, and the emphasis on very large scale projects that have characterized the practice of the subject in Holland ever since. And then there were his students, Willem de Sitter and Jan Oort, who shaped the study of cosmology and galaxy structure throughout the twentieth century. Jan Oort in particular discovered, along with the Swede Bertold Lindblad, the differential rotation of the Milky Way about its center. Oort applied Kapteyn's measurements to determine the direction and the distance to the center of the Galaxy and in his later years became driven by curiosity about that particularly singular location and the strange processes going on there. But first, Oort had to find a way to part the curtain of dust.

3

The New Physics

3.1   Revolution

While Kapteyn was occupied in Groningen measuring the positions of stars on photographic plates a revolution in physics was occurring elsewhere in Europe – a revolution that, over two decades, fundamentally changed the perception of the physical world that rested on the well-tested foundations of Newton's theory of dynamics and gravity and Maxwell's theory of electromagnetism. To say that these developments have had a profound impact on astronomy and astrophysics is a drastic understatement.

There were, in effect, two parallel revolutions. The first, following Max Planck, who in 1900 proposed that electromagnetic radiation came in discrete units of energy, was the quantum revolution. This permanently altered human perception of determinism and the fundamental limits of experimental precision and led to the concept of particle–wave dualism. The second revolution, the theory of relativity, forever changed the view of space and time as absolute entities and shook the foundations of Newtonian mechanics. The new theory of relativistic gravity, General Relativity, predicted phenomena that were discovered decades later, such as gravitational lenses, and, of particular relevance to this discussion, the existence of a new class of objects that are the consequence of gravitational collapse – black holes. It this second revolution that is most relevant to the discussion here.

It all began rather quietly in 1905 in Zurich where an obscure clerk in the patent office, Albert Einstein (Figure 3.1), wrote a fundamental paper dealing with the way in which the laws of particle dynamics and electromagnetism should be transformed between different reference frames moving at a constant relative velocity with respect to one another. As Kapteyn was the ultimate empiricist, Einstein was motivated more by matters of principle, and at that time there was

Figure 3.1 The young Albert Einstein. (Hebrew University of Jerusalem Albert Einstein Archives. Courtesy of AIP Emilio Segre Visual Archives.)

a well-recognized problem in principle: Newtonian dynamics is defined by Newton's three laws of mechanics: (1) free particles move with a constant velocity along a straight line; (2) the force on a particle is equal to its mass times acceleration; (3) the forces of action and reaction are equal and opposite. These laws remain the same when transformed between two frames in constant relative motion, that is, between inertial frames.

The mathematical rules for such a transformation are straightforward and intuitive: for the two frames the spatial coordinate in the direction of motion differs by a term Vt where V is the velocity and t is the time. Moreover the time is the same in both frames, which is to say that all clocks tick at the same rate; there is an absolute time. More generally we can say that all intervals in space and time remain the same. These are the so-called Galilean transformations that single out inertial frames as special. There is an infinity of inertial frames, so Newton picked one frame as really special – absolute space. Acceleration with respect to this absolute space is resisted, and that resistance (described by $F = ma$) is inertia.

But the problem in principle that confronted physicists in the beginning of the twentieth century was that Maxwell's equations, describing the relation of electric and magnetic fields to charges and currents, did not remain the same with this sort of transformation (these equations are not invariant under Galilean transformations). Maxwell's equations in a vacuum (no charges or currents) predict the existence of electromagnetic radiation, including visible light, all traveling at the velocity of light, c. But, in a Galilean transformation, the velocity between the two frames is added or subtracted from the velocity of light so it becomes $c' = c \pm V$.

When he wrote down his equations in 1873, James Maxwell proposed that the radiation is a wave disturbance in a hypothetical medium, the aether, and the equations were strictly valid only in a frame at rest with respect to this medium. Perhaps the frame of the aether should be identified with Newton's absolute space. But then, the velocity of light should differ in frames moving with respect to that medium; for example, the motion of the Earth about the Sun (18 km/s) ought to be detectable as a variation of the speed of light when measured in different directions on Earth and at different times of year as the direction of the Earth's motion varied. In the 1880s the American physicists Albert Michelson and Edward Morley tested this prediction with an interferometer – an instrument extremely sensitive to variations in the speed of light with direction. They found no such effect resulting from the motion of the Earth about the Sun.

In 1900, Henrik Lorentz, a physicist at Leiden University, discovered (following earlier work by the Irish physicist George FitzGerald) a prescription that would work – that would explain the results of the Michelson–Morley experiment. In this transformation, distance intervals contracted in the direction of motion with respect to the aether in just such a way as to keep the speed of light apparently constant, that is, by a factor of $\gamma = 1/\sqrt{1 - V^2/c^2}$. So the transformed distance intervals in a moving frame appear to shrink in an observer's frame (the aether frame) by a factor of $1/\gamma$. These "Lorentz transformations," later generalized by Joseph Larmor and Henri Poincaré, modified times intervals as well as length; time intervals in the moving frame appear to be longer or "dilated" to the stationary observer by a factor of γ. That is to say, a moving clock appears to tick more slowly.

Although this sort of transformation worked fine for electrodynamics, it failed for Newtonian mechanics; that is, Newtonian mechanics is not invariant to Lorentz transformations. So it was thought that the contraction of the length of the interferometer in the direction of motion was an actual shrinking of the material, supposedly due to the intermolecular electromagnetic forces within the material (shown by Larmor to be inadequate). The dilated time interval in the moving frame was considered to be a sort of "local" time; no one could imagine that clocks would really run at different rates in frames in relative motion. All of this was a "fix" to save the immobile aether.

The situation in 1905 can be summarized like this: Maxwell's equations maintained their form in frames in constant relative motion using a Lorentz transformation, while the older and simpler Galilean transformation worked fine for Newtonian dynamics. But this was exactly the problem that disturbed young Einstein: Why did these highly successful theories, Newtonian dynamics and Maxwell's electromagnetism – the foundations of physics at that time – transform in different ways between moving frames? It seemed to Einstein to be a very unnatural, asymmetric, and inelegant way for Nature to behave.

Einstein felt that the same transformation rules should hold for both sets of phenomena, and he valued this principle more than the exact form of Newton's laws. So he made the bold leap of postulating that the speed of light was constant in all frames, which singles out the Lorentz transformations as valid for mechanics as well as electrodynamics. But then Newtonian dynamics had to be modified – the first modified Newtonian dynamics; this modification became particularly important at velocities near the speed of light (300 000 km/s), so at the rather lower velocities accessible in the laboratory the classical tests of Newtonian dynamics should be satisfied to high precision. And time was not absolute; clocks in different inertial frames really do tick at different rates. This was the concept that was most difficult for traditional physicists to grasp.

One consequence of the Einsteinian modified dynamics is that the mass of a particle increases with its velocity as γ, which is to say that as the velocity approaches the speed of light the mass approaches infinity. Another famous consequence is the equivalence of mass and energy, $E = Mc^2$. By now, more than 100 years later, both of these predictions have been well tested and confirmed to high precision. Modern versions of the Michelson–Morley experiment have confirmed the isotropy of the speed of light (the fact that it is the same in all directions) to better than one part in 10^{22}. Einstein's assumption of the validity of Lorentz invariance, for particle dynamics as well as electromagnetism, remains experimentally unshaken.

In a single stroke, Einstein overturned the sacred principles of absolute space and universal time as well as the concept of an invisible aether for transporting electromagnetic waves. It was a tectonic shift in human thinking on time and space. But his 1905 theory of relativity was special to frames in constant relative motion to each other. He immediately set upon the more difficult task of generalizing this theory to include accelerated motion, and, consequently, gravitation.

3.2   The Power of Principle

It took Einstein ten years to work out the complete General Theory of Relativity – years in which he went from his lowly job as a patent clerk in Zurich to professor and director of Kaiser Wilhelm Institute for Physics in Berlin. The

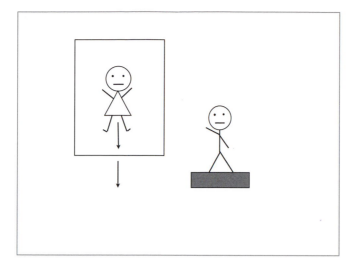

Figure 3.2 Sally, on the left, is in a closed elevator that is freely falling in the uniform gravitational field of Earth. Because she is falling at the same rate, she does not feel the force of gravity; she is weightless. Charley on the right is on a platform at rest with respect to Earth; he sees the falling elevator and interprets this as motion in a gravitational field. Both viewpoints are equally valid (until the elevator hits the ground).

theory, although technically complicated, is based on a few simple principles, the most important of which is the Equivalence Principle. It had been known since the time of Galileo that objects of different weight and different composition fall at the same rate: in a vacuum heavier objects do not fall faster than lighter objects, and objects made of iron do not fall faster than objects made of wood. Einstein was impressed with this experimental fact and elevated it to the level of a principle.

There are several ways of phrasing this principle. The broadest and weakest statement of the principle is that the rate at which an object falls in a gravitational field is independent of its internal structure or composition. A stronger way of putting it is to say that an observer, Sally, in a freely falling frame on Earth – a closed elevator, for example (Figure 3.2), in which the cable has broken – has no way of knowing whether or not she is freely falling in the gravitational field of a massive object or floating in interstellar space far from any star. There is no physical experiment whatsoever that she can do that can make that distinction; all the laws of physics are the same in every inertial frame. This is a strong statement of the principle and that taken by Einstein.

Such a freely falling elevator constitutes a "local inertial frame" even in the presence of a gravitational field. That means it is always possible to find a local inertial frame in which the paths of particles are straight lines, as Newton's first law specifies they should be (although in environments where the gravitational field changes greatly over small distances it may be necessary to take a very small elevator). But from the point of view of an observer, Charlie, on the stationary Earth watching the elevator fall, the particle paths inside the elevators are not straight lines; they are curved. Imagine a ball tossed from one side of the elevator to the other. From inside the elevator it appears to move on a straight line, but from outside its path is a parabola; it shares the accelerating motion of the elevator downward.

We may transform the laws of motion between the two frames by a change of coordinates – one coordinate system attached to the falling local inertial frame (Sally's frame) and a second coordinate system attached to the frame that is stationary with respect to the Earth (Charlie's frame). But we would like for the form of the equations of motion to be the same in the two frames; that is to say, the theory should be independent of such a general change of coordinates. For the more mathematically inclined reader, this is the principle of General Covariance, which follows from the Equivalence Principle but, mathematically, is more powerful. It is clearly more general than, but includes, the Lorentz transformation.

The Equivalence Principle can also be stated as an equivalence between inertial force (as in $F = ma$) and the gravity force; one can be shifted to another by a change of frame. This has an immediate predictive consequence. It can be shown in a straightforward way that this equivalence of inertia and gravity predicts that clocks deep in a gravitational field appear to run slower for an observer farther out in the field; that is to say, any clocks of an observer near the surface of Earth run slower from the point of view of an observer 100 km above the surface of Earth. An additional prediction immediately follows and that is the phenomenon of gravitational redshift: light emerging from the observer at the surface of the Earth appears shifted to longer wavelengths to the observer 100 km above Earth.

This equivalence of gravity and inertia has the consequence that the elements describing the underlying geometry enter the equations of motion for the particle – gravity and inertia are "geometrized." Particles (and light rays) follow special curves on the background geometry – geodesics – independent of their mass or composition. Geodesics are the shortest path between two points in a space, in this case a curved four-dimensional space–time. On the two-dimensional surface of a sphere geodesics would be great circles – like the curves of longitude on the surface of the Earth.

But what causes the structure of space–time to be curved? Einstein supposed that it was the mass-energy content of that space–time (recall mass and energy are equivalent), and in this he drew on another principle: space–time acts on the mass–energy constituents of the Universe, but mass–energy shapes space–time;

nothing (such as absolute space) acts on something else (such as the motion of particles) without being acted on. This is an aspect of Mach's principle, named after Ernst Mach, the Austrian physicist who had a great influence on Einstein's thinking.

The precise relationship between that mass–energy density and the curvature of space was the sticking point, and Einstein wrestled several years with this problem. But finally, with the help of his mathematician friend Marcel Grossman, he saw that the solution lay in a rich mathematical structure invented sixty years before by the German mathematician Bernhard Riemann. Riemann devised a method of describing the curvature of space in an arbitrary number of dimensions; basically at every point there is a collection a numbers, a tensor, which described the properties of the underlying space. These can be reduced to a single number, the scalar curvature – positive, negative, or zero – of that space (with zero corresponding to Euclidian flat space). This number is independent of the actual coordinate system; it is a coordinate-independent means of describing curvature, exactly what is needed for a coordinate-independent theory. With this breakthrough in realization, Einstein wrote down his famous field equations for General Relativity, overall an incredible achievement of human intellect and one largely without empirical input (other than the requirement that the law of attraction should be that of Newton far from a massive object).

The complete theory, published in 1916, had immediate observational consequences. The first was that an explanation was provided for a well- and long known problem of planetary motion – the precession of the orbit of Mercury. The orbit of Mercury is, like all planetary orbits, an ellipse. According to Newton, in the perfect inverse square gravitational field of a point mass the major and minor axes of the ellipse should be frozen in space but for no planetary orbits is this the case; for all planets, the orbital ellipses precess and slowly rotate about the Sun with respect to the distant, "fixed" stars. Most of this could be understood in the context of Newtonian gravity as being due to the small influence of the other planets; that is to say, the gravitational field was not that of a perfect point mass. But with Mercury, when all of these additional planetary perturbations were included, there still remained an unexplained 40 arc seconds per century of precession (an arc second is quite a small angle; a two kilometer crater on the surface of the Moon would subtend about one arc second as seen from Earth). This problem had been appreciated by the French mathematician Urbain Le Verrier, who in 1846 had predicted the existence of Neptune on the basis of such anomalies in the motion of Uranus. He tried the same with Mercury, postulating the presence of several undetected planets closer to the Sun. When looked for they were not found. This led the Canadian-American astronomer Simon Newcomb (1882) to propose a breakdown of Newtonian gravity near the Sun, but by the early twentieth century, von Seeliger had provided a model for a distribution of small particles (presumably those causing the Zodiacal light) that did the trick without modifying gravity. This was generally considered to be an acceptable astrophysical model for the phenomenon.

And then came Einstein with General Relativity, which provided a perfect explanation for the extra precession of Mercury's orbit. After Einstein's theoretical explanation, a true modification of gravity, the Seeliger model seemed contrived and artificial. But the significant point is that Einstein was in no way influenced by the problem of the anomalous precession of Mercury in developing his theory. He was driven by pure principle: the desire to create a relativistic theory of gravity that embodies strong equivalence and ideas of Mach.

The second immediate consequence of General Relativity is a true prediction of a previously undiscovered phenomenon: that is the bending of the path of light in a strong gravitational field. Part of the bending is a direct consequence of the Equivalence Principle. Looking back at the falling elevator, if Sally is to see a light ray from one side of the elevator to the other and not be tipped off that she is falling in a gravitational field, then its path must be a straight line. But if this is true for Sally, then Charlie must see a curved path. This again is a consequence of the fact that all particles, including light (photons), follow geodesics of the underlying geometry (although photons and other particles moving near the speed of light follow special geodesics known as null geodesics – zero distance in the four-dimensional space–time).

Assuming that light has mass and responds to a gravitational field, then such a prediction can also be made for Newtonian gravity, as had been realized first by Newton (1704). A detailed calculation was published by Johan Georg von Soldner in 1801. Because of this deflection of light from a straight-line path the position of a background star passing near the edge of the Sun would appear to be shifted by about 0.8 arc seconds. Einstein in 1911 calculated what this should be, based on the Equivalence Principle, and came up with the same value. With the development of the full theory of General Relativity and the use of Riemannian geometry, he later realized that the actual value would be exactly twice as large or 1.75 arc seconds.

3.3   Arthur Eddington Boosts Einstein to World Acclaim

Arthur Eddington (Figure 3.3), the famous British astrophysicist, was not generally beloved during the First World War. He was a very high profile Quaker and a pacifist and refused to participate in the wholesale slaughter going on across the channel. He applied for, and received, conscientious objector status and managed to avoid prison owing to the intervention of influential friends.

Figure 3.3 Arthur Eddington, an early supporter of General Relativity and promotor of Einstein. (AIP Emilio Segre Visual Archives. Gift of Subrahmanyan Chandrasekhar.)

Eddington, like Kapteyn, also believed that science should be above national conflicts and so wished to maintain contact with German scientists. This was certainly not an easy task for an individual of dubious loyalty during a desperate struggle but was accomplished largely through the efforts of Willem de Sitter, Kapteyn's student who had become director of the observatory at Leiden. de Sitter, as a citizen of a neutral country, could correspond with both Einstein and Eddington, and through this correspondence Eddington developed an understanding of General Relativity and an enormous respect for Einstein and his accomplishment. He became the first prominent supporter of the theory and, in effect, its public relations agent in the West.

After the war Eddington was determined to test the theory observationally, in particular the prediction of light deflection in the gravitational field of the Sun. Only during an eclipse was it possible to photograph stars near the limb (edge) of the Sun and measure their positions, so Eddington organized and participated in the most famous eclipse expedition of all: that of May 1919 to the island of Principe off the coast of Africa. The results, within a fairly large error range, were consistent with prediction of General Relativity.

Einstein became an overnight celebrity. Justifiably, for to proceed from simple and clearly stated principles, through a complicated formalism, to a precise numerical prediction on light deflection about the Sun, a previously unobserved phenomenon that the informed public could understand, was a truly remarkable achievement. But another prediction was lurking in the first exact solution to Einstein's field equations – a prediction that would be confirmed only sixty to seventy years later.

3.4   The Solution from the Trenches

Karl Schwarzschild (Figure 3.4) grew up in the late nineteenth century in a comfortable Jewish family in Frankfurt and showed an early interest in astronomy. This interest was not only in the practical aspects of the subject; he also excelled in mathematics and published solutions for the orbits of double stars while he was still in high school. Schwarzschild received his doctorate at Munich under the direction of von Seeliger with a dissertation that was typical of his skills: an application of Poincaré's theory of stable configurations of rotating spheroids to astronomical objects such as moons of planets – a confrontation of a highly mathematical construction with observations of real objects.

By 1914, having by age forty advanced to the directorship of the Potsdam Observatory, he already had a highly successful career behind him. At the outbreak of

Figure 3.4 Karl Schwarzschild. (AIP Emilio Segre Visual Archives. Courtesy Martin Schwarzschild.)

war he was caught up in the nationalistic fervor that swept through all European states and enlisted in the Imperial Army. By way of Belgium and France, he finally arrived at the eastern front in Russia, where he was occupied calculating the trajectories of artillery shells. There, he became ill with an autoimmune skin disease from which he perished in 1916, but in the months before his death, he made a profound contribution to physics: he worked out the first exact solution to the very complicated field equations of Einstein – the relativistic gravity field in the vicinity of a mass concentrated at a single point.

In the Schwarzschild solution for the gravity field of a point mass, there is a special radius given by

$$R_s = 2GM/c^2$$

where G is the Newtonian gravitational constant, M is the mass, and c, as earlier, is the speed of light. At distances large compared to this "Schwarzschild radius" the gravitational acceleration due to a point mass becomes identical to Newtonian inverse square attraction ($a = GM/r^2$); at closer distances small deviations from Newton's law appear, such as the deviation from inverse square that causes the anomalous precession of the orbit of Mercury. Very near the critical radius there are dramatic differences, and from distances smaller than R_s no particle, not even light, can escape. There is, in effect, a horizon at R_s. An outside observer can have no knowledge of events occurring within this radius; it is a one-way surface – sort of an ultimate Hotel California: you can check in but you can never, never check out.

That such a radius could exist was already suggested in the context of Newtonian gravity because this is the classical radius where the escape velocity becomes equal to the speed of light. In 1783 the British geologist John Michell pointed out that if an object with the density of the Sun were actually 500 times larger than the Sun, which is to say, if the mass were 100 million times that of the Sun, then light leaving the object would return to its surface (this actually is the mass range relevant to the mass range of black holes in the centers of active galaxies). The same idea was discussed by the French mathematician Pierre-Simon Laplace a few years later. The point is that, within the context of Newtonian gravity, if an object lies within this radius, then no light can escape from its surface. From now on I refer to such hypothetical constructs as *black holes*, even though the term was invented fifty years later by John Wheeler.

The critical radius is proportional to the mass, but the average density within that critical radius is proportional to mass divided by the volume or the third power of the radius; thus the density is proportional to the inverse square of the mass. More massive black holes have a lower average density within their Schwarzschild radius. So Michell's 100 million solar mass black hole would have an average density of that of the Sun but would have a Schwarzschild radius of twice the distance of the Earth from the Sun. A black hole with the mass of the Sun would have a radius of about 3 kilometers but a mean density of about 2×10^{16} g/cm^3, in excess of the density of atomic nuclei. A handy formula for remembering the magnitude of the Schwarzschild radius is

$$R_s = 3M_* \, \text{km}$$

where M_* is the mass of the condensed object in units of the mass of the Sun (2×10^{30} kg). The actual radius of the Sun is 200 000 times larger, so the Sun is well outside of its Schwarzschild radius; it is certainly not a relativistic object.

Far from the Schwarzschild radius, particle orbits are essentially as in Newtonian gravity (they follow Kepler's laws), but within a few Schwarzschild radii of the black hole, particles move quite differently. With Newton, a particle in a circular orbit is always stable; this is guaranteed by the perfect inverse square law. But in General Relativity, a circular orbit within three Scwharzschild radii is unstable – the particle will spiral into the Schwarzschild radius and never be seen again. Surprisingly, circular orbits are also possible for light rays, or photons. At exactly 1.5 R_s, there is an unstable circular orbit for a photon: a slight perturbation inward, and the photon would spiral in; a slight perturbation outward, it would spiral out.

The clocks of an observer falling through the Schwarzschild radius (Sally) and one far away at a safe distance (Charlie) tick at very different rates. Sally finds that she passes the Schwarzschild radius in a finite time, but Charlie finds that it takes an infinite time for the unfortunate Sally to reach the Schwarzschild surface; Charlie observes that light being emitted by the infalling Sally becomes more and more redshifted (the gravitational redshift) until finally she can no longer be seen.

However, before Sally slips behind the Schwarzschild radius she runs the risk of being torn apart by the tidal force from the black hole. It is true that when she is freely falling in a *uniform* gravitational field, she cannot detect the fact that she is in a gravitational field at all; this is the Equivalence Principle. But near the black hole, the gravitational field is not uniform; it can vary greatly over small distances, even the 175 centimeters of Sally's length. The force pulling down on her feet can be much larger than the force pulling down on her head; she can be torn apart by tides. Now the tide at the Schwarzschild radius is proportional to the mass of the black hole and the inverse cube of the Schwarzschild radius ($1/R_s^3$). But the Schwarzschild radius is also proportional to the mass of the black hole, so the tidal force varies as the inverse square of the black hole mass. Smaller black holes are more dangerous than massive black holes. For a stellar mass black hole, Sally would perish long before she reached the horizon, but for a supermassive black hole, greater than one million solar masses with a Schwarzschild radius of three million kilometers, she would pass through the horizon without noticing much (this, as we shall see, is actually relevant to stars that encounter a black hole).

But then, after passing the horizon, she must confront the strangest aspect of the Schwarzschild's solution: the mathematical and physical singularity at zero radius. It was first thought that there was a singularity at the Schwarzschild radius as well. But in 1924, Eddington demonstrated that this was not a real physical singularity but a coordinate singularity – it can transformed away by a change of coordinates. This is not true of the singularity at $R = 0$, however. All physical quantities, such as the density, the curvature, the acceleration, and the tidal force, became infinite at this point. Moreover, any matter or radiation energy passing through R_s will inevitably reach the singularity in a finite time in its own frame; once inside the Schwarzschild radius, the singularity cannot be avoided (Sally is doomed). Primarily for this reason, for decades, many scientists (including Einstein himself) rejected the idea of a black hole; it all seemed too unphysical. Now it is thought that new physics may well intervene near the singularity – quantum gravity effects – but whatever happens there it is somewhat irrelevant because the singularity is closed off to the outside Universe by the horizon at the Schwarzschild radius.

In any case the theoretical basis for black holes existed for fifty years before it was applied to actual astronomical objects. It was the crisis provoked by the discovery of extreme astronomical sources that provoked reconsideration of such bizarre constructs. But before we consider this, I resume the journey inward to the center of the Galaxy.

4

Parting the Veil with Radio Astronomy

4.1 Kapteyn's Famous Student

Within five years of Kapteyn's death in 1922, his view of the Universe had been overturned. By 1930 it was generally accepted that the Milky Way Galaxy was one of many such systems supported against its own gravity by rotation, as had been supposed by Kant. This altered perception of the Galaxy was largely due to the work of the Swedish astronomer Bertil Lindblad and to Kapteyn's own student, Jan Oort (born 1900).

Oort (Figure 4.1) received his doctoral degree in Groningen in 1921 (the doctoral is roughly equivalent to a master's degree). After spending several years at Yale Observatory he was invited back to Leiden by de Sitter, and there he spent the remainder of his long active career.

For his PhD dissertation, defended in Groningen in 1926, Oort had studied the motions of halo stars, objects belonging to the Galaxy but distributed, as the globular clusters, in a large spheroidal halo surrounding the disk of the Milky Way. He was perplexed by the high velocities of these stars and the extremely skewed velocity distribution: they all seemed to be moving in the same direction. Then, in 1927, Lindblad demonstrated that Kapteyn's two star streams could be understood in terms of differential rotation of the Milky Way disk; which is to say, the system rotates not like a solid body, as a phonograph record, but the stars at any radial distance from the center revolved about the center at a rate determined by the Galactic gravitational force at that point. This opened Oort's eyes to the correct interpretation of his own result: the very high velocities and skewed velocity distribution of halo stars could be understood if they did not generally participate in this rotational motion, if the disk including the Sun were rotating through this spheroidal system so these stars appeared as a wind blowing from the direction of rotation.

Figure 4.1 Young Jan Oort about the time when he was a student in Groningen. (Copyright by Leiden Observatory.)

But Oort went further. Using Kapteyn's data on stellar positions and motions, he noticed regularities in the kinematics of the stars near the sun – regularities both in the radial velocity, the velocity directed along the line of sight from the Earth to the star that can be measured spectroscopically through the Doppler shift, and in the proper motion, the apparent motion on the sky which is proportional to the velocity perpendicular to the line of sight. When, for a number of stars, either of these two velocity components are plotted against the angle between the star and the direction of the Galactic Center, a systematic pattern emerges: the points seem to trace out sinusoidal curves. Oort realized that such a pattern for nearby stars could be understood entirely in terms of differential rotation of the Galaxy.

This is illustrated in Figures 4.2 and 4.3 for the case of radial velocity. Here we see that for a star at zero degrees – in the direction of the center – there is no component of the star's rotation velocity about the center in the direction of the Sun (apart from the random velocity with respect to the center which, averaged over a number of stars, should be zero). Also at 90 degrees both the Sun and the star are at the same radial distance from the center and moving in the same direction with about the same velocity. Therefore, the average radial velocity of these stars would also be zero. But at 45 degrees, there is a significant component

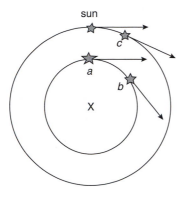

Figure 4.2 From the position of the Sun, star *a* at a longitude of zero degrees has no component of velocity along the line of sight; star *c* at 90 degrees shares the rotation velocity of the Sun and therefore also has no net component of line of sight velocity; but star *b* at 45 degrees has a substantial net component of velocity directed away from the Sun. Thus Galactic rotation gives rise to the systematic relationship between the radial velocity of stars near the Sun and their Galactic longitude shown in Figure 4.3.

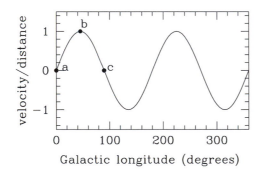

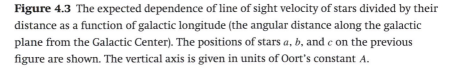

Figure 4.3 The expected dependence of line of sight velocity of stars divided by their distance as a function of galactic longitude (the angular distance along the galactic plane from the Galactic Center). The positions of stars *a*, *b*, and *c* on the previous figure are shown. The vertical axis is given in units of Oort's constant *A*.

of a star's velocity away from the Sun, so the result is the double sine shown in Figure 4.3. The formula for the average radial velocity derived by Oort is

$$V_r = A\,d\,\sin(2l)$$

where *l* is the angular distance in the Galactic plane between the star and the center (which became designated as Galactic longitude), *d* is the distance to a particular star, and *A* is a constant – the famous Oort's *A* constant. There is

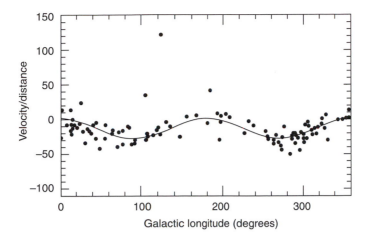

Figure 4.4 The tangential velocity for Cepheids within one kiloparsed. From recent data taken with the Hipparcos satellite. The vertical axis is essentially the tangential velocity divided by the distance to the particular star (with corrections made for solar motion). Oort's constant A is one-half the vertical peak-to-peak distance and the constant B is the vertical offset of the curve from zero. Notice that $B \approx -A$, which is the case for a flat rotation curve. This is from an analysis by Michael Feast and Patricia Whitelock.

a similar formula for proper motion but with a vertical offset, Oort's constant B. A plot of proper motion vs. longitude, based on modern data, is shown in Figure 4.4.

The constants A and B depend on the rotation velocity of the Galaxy at the position of the Sun, the distance to the center of the Galaxy, and the rate at which the rotation velocity varies with distance to the center. Oort estimated numerical values for A and B by making such longitude–velocity plots. Then, from his own work on halo stars, he took the rotation velocity of the Galaxy at the position of the Sun to be about 260 km/s and, using his formulae, estimated the distance to the center to be about 20 000 light years. These estimates for rotation velocity and distance to the center are within 20% of their currently determined values, but the distance is significantly less than Shapley's estimate based on the system of globular clusters (due to Shapley's neglect of interstellar absorption). Moreover, from the phase of the double sine curve (the angle at which the radial velocity is zero) Oort was able to determine the direction of the center of the Galaxy: roughly in the direction of the constellation of Sagittarius.

He went further. He pointed out that if the entire Milky Way had the density of stars observed locally, then it could not possibly provide a rotation velocity of 260 km/s; there was just not enough Galactic mass inside the position of the

Sun to provide the necessary gravitational force. He concluded that the density must be much higher in the direction of the center than it is locally. But that is not what his old teacher Kapteyn had observed. Kapteyn saw the density of stars falling off uniformly in all directions in the Galactic plane. Thus, there must be obscuring matter generally distributed in the plane of the Milky Way (Oort called it dark matter), much more than Kapteyn or Shapley had thought. It was the presence of this obscuring, non-luminous matter, now known to be small particles of interstellar dust, that caused the apparent density of stars to decline uniformly in all directions, even toward the center. And this had led Kapteyn to his erroneous conclusion that the Sun was near the center of the Milky Way system.

This was a remarkable achievement for the young astronomer. He had totally destroyed the cosmology of his master and he did it by looking at very nearby stars. In the words of Woodruff Sullivan, the author of *A History of Radio Astronomy*, it was as though Oort had deduced the existence, direction, and distance to Amsterdam by observing the motion of individual people in the marketplace in Leiden. By 1930, due in no small measure to Oort, there was a completely different dominant world view than had existed ten years earlier. The Universe consists of galaxies and the Milky Way is only one of these. The Sun is an obscure star at the outskirts of this average galaxy. Moreover, the galaxies are rushing away from one another with a velocity proportional to their separation, as observed by Edwin Hubble in 1930. Hubble found that the spectral lines of distant galaxies are shifted to the red – the redshift – by an amount proportional to their distance. Because we are not in a special position this means any observer in any galaxy will see the same. The Universe is truly vast and uniformly expanding; this was totally consistent with the second exact solution of Einstein's equations worked out by the young Russian mathematician Alexander Friedmann in 1920. Friedmann had, on the basis of General Relativity, predicted a nonstatic Universe before it was discovered.

In spite of his remarkable achievement, Oort was not satisfied; he wanted to know more about the Milky Way. Presumably our Galaxy had the conspicuous bright spiral arms seen in other such galaxies as well as a dense bright central region where the density of stars was much greater. How could he observe and map this structure given the presence of the obscuring dust? How could he determine the rotation law, and hence the mass distribution of the entire galaxy – right into the center? This was the problem: how to part the veil.

4.2   Radio Astronomy and the 21-cm Line

By the outbreak of World War II, Oort, having turned down offers of prestigious positions at Harvard and Columbia in the United States, had become extraordinary professor at Leiden. Following the German invasion of the Netherlands in 1940, he refused to cooperate with the occupying authorities in the nazification of Leiden University; his continued presence there became risky, so he resigned his professorship and moved his family to the relative obscurity of a small village, Hulshorst, about 100 kilometers east of Leiden. However, he maintained contact with colleagues in Leiden and in Utrecht via bicycle, occasionally giving seminars and suggesting projects to various staff members.

In 1944 Oort became excited by the prospect of observing the Galaxy at radio wavelengths. Earlier, in 1933, Karl Jansky, working at Bell Labs in New Jersey, had discovered celestial radio emission apparently from the Galaxy because it peaked when the galactic plane was overhead. This motivated Grote Reber, a very gifted American amateur astronomer, to construct his own radio telescope, a partially steerable parabolic dish of 9 meters operating at a wavelength of 10 meters, and by 1943 he had mapped the Galaxy with a resolution of 10 degrees. Reber discovered, in addition to the smooth radiation from the galactic plane, several discrete or point-like sources of radio waves (at least, point-like to Reber's telescope beam).

There are two sorts of electromagnetic radiation that can be detected from astronomical objects. There is radiation at all frequencies, or continuum radiation. Any object with finite temperature emits continuum or black body radiation – the Sun, for example with a temperature of around 6000 degrees Kelvin emits visible radiation peaking at a wavelength of 4800 angstroms detectable to the human eye as yellow light. Jansky and Reber observed continuum radiation at radio wavelengths, and it was originally thought that this was just such thermal emission from all of the stars along the line of sight through the Galaxy (much later shown to be nonthermal emission from relativistic electrons spinning in the Galactic magnetic field).

Then there is line radiation – radiation at a particular frequency emitted by a particular substance such as hydrogen. The spectrum of the Sun and other stars contains many such spectral lines characteristic of hydrogen, helium, and other trace elements. An advantage of a spectral line is that the motion of the source, toward or away from the observer, can be measured by its Doppler shift to the blue or to the red.

This was the prospect that excited Oort. He realized that if there were such a spectral line at radio wavelengths, then it would not be obscured by the interstellar dust in the galactic plane – the dust that made optical observations of stars more distant than 1000 light years or so impossible. That is because the dust particles obscure most effectively radiation with wavelength comparable to the size of the dust particles. The dust particles obscuring the visible light were therefore tiny and would impose no barrier for radio waves. If only there were such a line at radio wavelengths from some abundant element such as hydrogen, then one could look across the Galaxy and map the distribution and motion of the gas. One could measure the rotation curve of the Galaxy and determine the distribution of force and mass. One could peer all the way to the center of the Galaxy.

This was the project he suggested to Henk van de Hulst, a bright student at Utrecht University: look into the possibility of a spectral line from neutral hydrogen at radio wavelengths. Van de Hulst threw himself into the project. He first considered high recombination lines of hydrogen. Recall that a hydrogen atom consists of a negatively charged electron orbiting a positively charged proton, but (an aspect of the quantum revolution) only at discrete energy levels or distances. These discrete energy levels get closer and closer together as the distances get larger until finally they merge into a continuum of states corresponding to complete ionization of the hydrogen atom. So, although the transition between the second and first excited levels corresponds to a line in the visual part of the spectrum, the famous Balmer α line, the wavelength of radiation emitted during transitions between high levels will be at radio wavelengths. Van de Hulst decided (erroneously) that, due to gas pressure, these lines would be too broad to be detected. He then considered a so-called hyperfine transition in the hydrogen atom. When the spin of the electron is parallel to that of the photon, the energy of the configuration is slightly higher than when it is antiparallel. A transition between these two states would emit (or absorb) a photon at a wavelength of 21 centimeters. Van de Hulst concluded that this line from the interstellar medium might be detectable if the sensitivity radio receivers at that time could be improved by a factor of 100.

After the war, Oort was invited back to Leiden to become director of the observatory and began efforts to construct receivers and antennae in order to detect the 21-cm line. Given the postwar conditions in Holland, progress was slow; there was a considerable push from Oort and the astronomical community, but the resources, material and human, were very limited. At the same time, a serious effort was underway at Harvard, where a young graduate student, Harold Ewen, and his advisor, the eminent physicist Edward Purcell, had become excited about van de Hulst's proposal and the prospect of detecting the 21-cm line – an excitement reinforced by the Soviet astrophysicist, Iosif Shklovsky, who published an important paper in 1949 on the possibility of detection (in the Russian language *Astronomical Journal*). Shklovsky had read about van de Hulst's proposal and worked out the signal strength expected from the interstellar medium: the line should be easily detectable.

In 1951, following a tradition established by Kapteyn in which the best Dutch students were sent abroad, van de Hulst took a position as visiting professor at

Harvard, where he heard about the efforts in the physics department. Very shortly after his arrival Ewen and Purcell detected the 21-cm line from the plane of the Milky Way. In the best spirit of scientific openness and international cooperation, Purcell and Ewen shared their discovery with van de Hulst and described the details of their radio receiver. Van de Hulst reported this discovery to Oort, who at that point was working with a young radio engineer, C.A. Muller. Within seven weeks, using an old 7.5-meter discarded German radar dish, Oort and Muller confirmed the Harvard result (followed very closely by W.N. Christiansen and J.V. Hindman in Australia). The tool to explore the entire Galaxy without the veil of interstellar dust had been provided.

Oort wanted and got the Dutch science foundation to fund a new 25-meter telescope for the purpose of exploiting this tool for Galactic research. From 1954 to 1956 this dish-shaped antenna was constructed in Dwingeloo in the relatively unpopulated and radio-quiet province of Drente, about 50 kilometers south of Groningen. This was, for a some months, the largest radio telescope in the world, until the completion of the 76-meter telescope at Jodrell Bank in the United Kingdom, and Oort knew exactly how to use it.

The task was to map the motion and distribution of neutral hydrogen throughout the Milky Way, and the first step was to determine the rotation law for the Galaxy: how does the rotation velocity depend on radial distance from the center? This is fairly easy at radial distances less than that of the Sun because hydrogen gas along a given line of sight in the plane of the Milky Way will have the largest radial velocity toward or away from the Sun at that point where the line of sight is at its closest distance from the center (see Figure 4.5). It is at that point, the

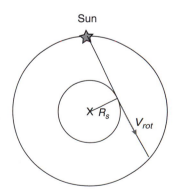

Figure 4.5 Measuring the rotation curve of the Galaxy. At a given Galactic longitude, the highest velocity 21-cm line emission is observed where the line of sight reaches the nearest point to the center, the subcentral point. At this point the rotation velocity lies entirely along the line of sight.

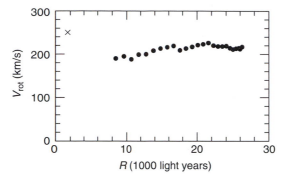

Figure 4.6 The rotation curve of the Milky Way in neutral hydrogen (Kwee, Muller, and Westerhout, 1954). The X shows the rotation velocity near the Galactic Center implied by the observations of Rougoor and Oort.

sub-central point, where the entire rotation velocity is directed away from (or toward) the Sun. Then it is a matter of simple geometry to determine the rotation velocity at any radius within that of the sun. Given the rotation velocity law for the Galaxy, there is a relation between the radial velocity and the distance for a particular emission feature in 21-cm emission, with a twofold ambiguity about the subcentral point. So one can then map the distribution of gas about the Galaxy and hopefully see the spiral structure (with some uncertainty given the ambiguity). This is what Oort, with a number of students and in collaboration with Australians (to fill in the southern part of the Milky Way), did over the next ten years, and the results, for the rotation curve and the implied gas distribution, are shown in Figures 4.6 and 4.7.

The rotation curve of Kwee, Muller, and Westerhout from 1954 is particularly interesting. The Galactic parameters (the distance to the center of about 25 000 light years and rotation velocity at the position of the Sun near 220 km/s) are very near the currently accepted values, and, in form, the rotation curve agrees very closely with the modern determinations. Notice in particular that the rotation curve is flat out to the position of the Sun, showing no hint of a Keplerian decline. This was contrary to the preconception at the time that the rotational velocity should be declining in the outer regions.

With this rotation curve in hand, and therefore the radial distribution of gravitational force, it was possible to model the mass distribution in the Galaxy. This was done by Oort's student Maarten Schmidt, who in 1957 produced a mass model of the Galaxy shaped roughly like a pancake. Unsurprisingly, the mass density increases toward the center of the Galaxy. But what about the mass distribution in the center? What could be deduced from 21-cm line observations?

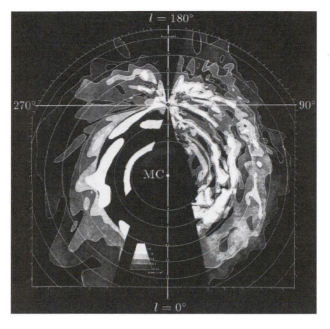

Figure 4.7 The Kerr–Westerhout neutral hydrogen map of the Galaxy. The lighter shades show the regions of higher gas densities. It was later shown by Butler Burton that large uncertainties in this implied gas density are introduced by small streaming motions associated with spiral structure.

4.3   Toward the Galactic Center

One of Oort's principal motivations for moving into radio astronomy was to observe the center of the Galaxy without the problem of obscuration. But there is a problem observing the Galactic Center from northern Europe. The center is in the constellation of Sagittarius 28 degrees below the celestial equator. This means that the center is at maximum about 10 degrees above the southern horizon and visible for only a few hours a day. The telescope at Dwingeloo must be pointed almost directly at the horizon to observe this most interesting region. Nonetheless, this was done in the late 1950s by G.W. (Wim) Rougoor, another in the succession of bright Oort students. Rougoor and Oort found very high velocities, up to 250 km/s within a few degrees corresponding to about 1500 light years, apparently due to rapid rotation of the gas (designated by the X in Figure 4.6). This implied an interior mass (interior to 1500 light years) of 6 billion solar masses or an average density of about 0.4 solar masses per cubic light year. This may not seem like a very large value (if spread out evenly it would be about 500 hydrogen atoms per cubic centimeter or a near perfect vacuum), but it is in fact more than ten times larger than the density of matter in the neighborhood of the Sun.

It became clear that the density of matter, presumably primarily old stars, was much higher in the inner 1000 light years of the Galaxy.

But could there be something more bizarre going on at the Galactic Center? In the 1950s a number of discrete sources of radio continuum emission were found around the sky. The most powerful source in a given constellation was given the designation "A," the second most powerful source, "B," and so forth; for example, the most powerful source in the constellation of Centaurus was Centaurus A. Some of these were apparently associated with objects in the Galaxy, such as Taurus A at the position of the Crab nebula, the remnant of a supernova seen by Chinese astronomers 1000 years ago. But others were associated with extragalactic sources such as Virgo A at the position of the massive elliptical galaxy M 87. Quite a number of sources had no identification at all; these were not concentrated in the galactic plane but spread uniformly around the sky and were called radio stars.

A particular galactic source, Sagittarius A (Sgr A), was discovered in 1954 by R.X. McGee and J.G. Bolton working with a 22-meter paraboloid dish (fixed in the ground) in Australia; they tentatively identified this source with the center of the Galaxy. Then, Rougoor and Oort identified the direction of the kinematic center of the Galaxy from their 21 cm line observations. They concluded in 1960 that "This position (of Sagittarius A) agrees so precisely with the direction of the galactic centre ... that this by itself makes it almost certain that Sgr A is situated at the centre of our Galaxy" (Oort and Rougoor 1960).

But Rougoor and Oort noticed something quite provocative. If we look back at Figure 4.5 we should notice that for a system in pure rotation, gas velocities at positive galactic longitude (0 to 180 degrees) and in the inner Galaxy (within the solar circle) should be directed away from the sun (positive radial velocity or redshifted). And at negative galactic longitude (180 to 360 degrees) also within the solar circle, the velocity of stars or gas should be toward the sun (negative radial velocities or blue shift). In particular, the velocity of gas directly toward the center should be near zero for pure rotation because it is moving at a right angle to our line of sight. This is not what Rougoor and Oort observed. The line profile of the 21-cm line directly toward the center, toward Sgr A, is shown in Figure 4.8. Usually the 21-cm line is seen in emission but, because of the presence of the bright continuum source at the Galactic Center, it appears here in absorption. There is a deep absorption feature at 0 km/s corresponding to all of the neutral hydrogen between the Sun and the Galactic Center. But there is also considerable absorption at velocities well away from zero; that is to say, there is considerable gas toward the center not moving on circular orbits.

A conspicuous absorption feature (first noticed by Hugo van Woerden along with Rougoor and Oort) occurs at a velocity of −53 km/s. The feature is moving

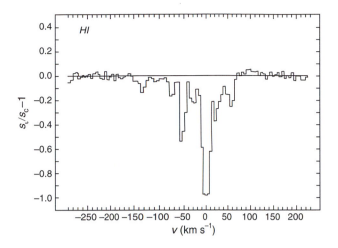

Figure 4.8 A modern observation of the 21-cm absorption line spectrum against Sgr A. The horizontal axis is the radial velocity in units of km/s and the vertical axis is the fraction of the continuum emission (actually in this case from the point-like source at the very center of Sgr A) absorbed. The deep feature near 0 km/s is all of the neutral hydrogen between the Sun and the center. The feature at −53 km/s moving toward us and hence away from the Galactic Center is the famous 3-kiloparsec arm. (From Liszt et al. 1983.)

toward the Sun and, because it is seen in absorption, away from the continuum source Sgr A. Beyond the position of Sgr A the same feature is seen in emission and can be traced to a galactic longitude of −20 degrees; this corresponds to a distance from the Galactic Center of 10 000 light years or three thousand parsecs – hence its designation as the 3-kiloparsec arm arm. If it is modeled as a partial ring (no comparable feature was seen at positive velocities), then it is rotating but also expanding, which is to say, moving away from the Galactic Center. This discovery came at a time (early 1960s) when there was increasing evidence of violent events in the nuclei of other galaxies, so this feature was taken by many to be evidence for similar events in the center of the Milky Way. As it turned out, this evidence was misleading, but it did fit into the picture that many galaxies, including our own, might show signs of energetic activity. But what is the source of this activity? What powers galactic nuclei in general?

4.4   Jan Oort and the Galaxy: An Assessment

At the end of the Second World War, when Oort became director of Leiden Observatory, a position that he held until his retirement in 1970, he moved into scientific management in the Netherlands and became a motivator of large projects such as the Westerbork Radio Synthesis Telescope, the first large interferometric array. He was a highly successful manager because he was driven by clear and achievable scientific goals and because of his highly developed social intelligence – he could smoothly navigate through the intricacies of the Dutch scientific community and funding agencies. He was an excellent judge of the abilities and shortcomings of his younger colleagues, and he knew which projects to assign to specific individuals. Students worked very hard for him because they respected his scientific judgment and they wanted to please him. He never gave orders, but he was very effective with a sort of gentle coercion. It was very difficult to say "no" to Jan Oort.

He moved the Netherlands into radio astronomy because he wanted to part the veil of dust that had been the curse of Kapteyn, and with radio astronomy, line and continuum, he discovered the Galactic Center with its wide hydrogen lines, a tracer of a central mass concentration, and its noncircular gas motions, interpreted at that time as an indicator of explosive or ejective activity.

Galactic nuclei had become a subject of intense interest in the 1960s. There were indications of activity going on in a number of galaxies, which seemed to strain the laws of physics. Was matter being created in and ejected from galactic nuclei? Was this the source of spiral structure? Even if these more bizarre suggestions were not true, how could such enormous radiant energy emerge from such a compact region? To answer these questions it made sense to observe the properties of the most nearby of galactic nuclei – that of the Milky Way.

5

The Violent Universe

5.1   A New Constituent of the Universe

The demands of war lead to technological advances that have consequences for pure research long after. The development of radar during World War II is one such example. The German radar dishes left behind in Holland at the end of the war enabled Oort to make a beginning for Dutch radio astronomy which, as we have seen, led to an understanding of the overall structure and kimenatics of the Milky Way Galaxy. Ewing and Purcell, the discoverers of the 21-cm line of neutral hydrogen, both had strong backgrounds in microwave radar technology, Ewing as a naval radar officer and Purcell as director of the MIT Radiation Laboratory. In the United Kingdom the young radio engineer Martin Ryle made a major contribution to the development of radar and thus to the British defense effort. But after the war, he applied these skills at Cambridge in the development of radio astronomy, in particular, the technique of radio interferometry.

In astronomy whenever a new wavelength window opens, major discoveries follow. This was certainly the case in the early 1950s, when crude radio telescopes began mapping the sky in continuum radiation and, in addition to smooth radiation from the Galaxy, discovered a number discrete sources scattered about the sky. Some were clearly Galactic in origin (they were in the plane of the Galaxy and associated with known objects such as supernova remnants) but others were more uniformly distributed outside of the Galactic plane. Some of these sources were apparently associated with galaxies, but many more were unidentified – the resolution was too poor to find a corresponding optical object with any certainty.

It was with this problem in mind that Ryle developed the technique of radio synthesis observations. Put simply, a radio interferometer is several antennae or dishes used together as a single telescope. The individual dishes are separated by meters or even kilometers, but the largest separation, or baseline, determines the resolution of the interferometer, resolution being the angular precision with which a source can be located on the sky (or more precisely, the minimum angular separation of two sources that can be resolved as two sources). For example, an interferometer with a 1-kilometer baseline has a resolution equivalent to that of single dish with a diameter of 1-kilometer; this would provide a resolving power of about 7 arc minutes, about one-fourth the diameter of the full moon, at a wavelength of 2 meters.

Earlier work had allowed several discrete radio sources to be identified with extragalactic objects. For example, Cygnus A was associated with a distant galaxy at a redshift of 0.057, corresponding to a recession velocity of 171 000 km/s or, from the Hubble relation, an incredible (at that time) distance of 600 million light years. From this it was clear that at least some of these radio sources were probably distant and luminous galaxies and possibly useful as probes of cosmology.

Ryle and his colleague Anthony Hewish understood that observations of radio sources would not become a proper astronomical and possibly cosmological tool until these sources could be identified with optical counterparts, and so constructed the Cambridge four-element interferometer with the largest baseline (east–west) of about half a kilometer. The first significant scientific result was the third Cambridge catalogue of radio sources (the 3C catalogue) published in 1959 and comprising 471 sources, many of which were undoubtedly extragalactic.

The Cambridge interferometer was more useful in compiling a complete list of discrete bright radio sources rather than as an actual tool for identification. The first actual new identification of a 3C radio source resulted from radio positioning observations began in 1960 by Thomas Mathews and Allan Sandage using the Caltech interferometer at Owens Valley with better than 10 arc seconds resolution. This was of 3C 48, identified with a faint star-like object surrounded by a wisp of nebulosity.

Second, in 1963 came 3C 273, identified by Cyril Hazard along with M.B. Macky and A.J. Shimmons at Parkes in Australia using the technique of lunar occultation (the source could be accurately positioned because of the Moon passing across); this source was identified with a relatively bright star-like object. But what were these objects? Did they belong to the Milky Way or were they at cosmological distances?

Maarten Schmidt (Figure 5.1) was Oort's student who had worked on mass models of the Galaxy following the initial determinations of the rotation curve from 21-cm line observations at Dwingeloo (Figure 4.6). After a postdoctoral position at Caltech, where he became involved in observations with the Palamar 200-inch telescope, he returned there permanently in 1959. Following the optical identification of 3C 273 (Figure 5.2) he obtained a visual spectrum of the object with a

Figure 5.1 Maarten Schmidt's discovery that star-like radio sources were at cosmological distances and therefore extremely powerful attracted wide public attention. He even made the cover of *TIME* magazine. (AIP Emilio Segre Visual Archives, John Irwin Slide Collection.)

Figure 5.2 An optical image of the quasar 3C 273. The object is unresolved but so bright that it masks the surrounding galaxy. The luminous jet, extending some 150 000 light years, is obvious on this print. (Image made by John Bahcall at the Hubble Space Telescope, NASA/ESA.)

number of lines corresponding to no element that could be immediately identified. After staring at the spectrum and thinking about it for some time, Schmidt
realized that these were the Balmer emission lines of recombining hydrogen but
redshifted by 47 000 km/s; with the Hubble law this would correspond to a distance of 2.4 billion light years, at that point one of the most distant object ever
detected. Now this was quite remarkable because the source is not particularly
dim; it is a 13th magnitude object, which means that, at its astounding distance
it must be intrinsically about twenty times more luminous than the entire Milky
Way galaxy. Yet, the object does not appear to be a galaxy – it looks like a star
with a faint jet-like extension to one side.

And 3C 273 was not a fluke. At about the same time Jesse Greenstein and
Thomas Mathews at Caltech measured a redshift of 0.367 for 3C 48, corresponding to a velocity of 30% of the speed of light. The Hubble law is certainly no longer
linear at these redshifts; cosmology must be taken into account. In the context of
the current LCDM cosmological model this would be at a luminosity distance of
6.2 billion light years (the luminosity distance is that used for converting apparent brightness into absolute luminosity). High redshifts and large luminosities
seemed to be the rule for these objects. But just as amazing – this extreme power is
arising in a very small volume. Just after optical identification, these "radio stars"
were found to vary by factors of two or three on timescales of days to months.
Such a significant variation cannot take place on a timescale shorter than that
required for light to traverse the object, so this means that the emission regions
have to be smaller than light months. That is to say, the radiant energy of a large
galaxy is emerging from a region a few times larger than the Solar System. This
discovery really did seem to strain the laws of physics.

The physics of these objects was discussed in 1964 in a classic paper by Greenstein and Schmidt. They considered three possible origins for the large redshifts:
(1) the objects are relatively close by but we are observing Doppler and not cosmological shifts; they are actually moving away from the sun with relativistic
velocities; (2) a substantial fraction of the redshift is gravitational; and (3) these
objects are really at the cosmological distances indicated by their redshifts.

They rejected the first possibility because of the absence of observable proper
motions – motion perpendicular to the line of sight. If these objects are moving
so fast, then why only away from us? They should have a substantial motion on
the plane of the sky as well that we should observe as proper motion unless they
are very far away. So this constraint would place the objects at extragalactic distances and their luminosity would be more comparable to that of galaxies than of
stars in any case (unless the objects are shot out of the Milky Way at high speed, a
possibility actually considered somewhat later). With respect to the second alternative, gravitational redshift, it is certainly possible that the emitting lines are

deep within a potential well; then, as discussed in Chapter 3, the light will be redshifted while climbing out. But Greenstein and Schmidt estimated a gravitational potential well necessary to produce the observed redshift would require a galaxy mass within a few light years and it was highly questionable if such a configuration would be stable (it should collapse to within its Schwarzschild radius on short order). In addition, the spectral lines must arise in an implausibly thin shell; otherwise, they would be broadened beyond recognition by the gravitational redshift.

It was the third possibility that Greenstein and Schmidt judged to be most likely: these actually are very distant extragalactic objects. But in this case there are also physical problems. The objects are incredibly luminous; they require a hitherto unimagined energy source. Moreover, given the small volume of the emission region, the energy density is extremely high. The radio emission was, by that point, identified with the synchrotron mechanism – relativistic electrons spiraling along magnetic field lines. But with such a high energy density of photons, another process should overwhelm the synchrotron: the relativistic electrons should scatter the photons up to X-ray energy and thereby very rapidly lose their energy. How could these sources be shining in synchrotron radiation?

Greenstein and Schmidt did not speculate on the source of the high luminosity but they did point out that an object of one billion solar masses well within a light year might also be within its Schwarzschild radius. They offered no mechanism for the emission from what should be a dark object.

Following the identification with star-like objects these sources were called quasi-stellar radio sources, and quickly thereafter, in a masterstroke of scientific terminology, "quasars" (invented by astrophysicist Hong-Yee Chiu in 1964). After the realization, by Allan Sandage, that there are many more such objects that are not powerful radio emitters (they are radio quiet), the more general term QSO or quasi-stellar objects came into use. The physical problems associated with such high luminosities and energy densities led some to suggest that the redshifts were intrinsic – not due to cosmological expansion but some new physics of undetermined nature perhaps connected with creation of new matter. In particular, Halton Arp supported this point of view on the basis of apparent positional associations between QSOs and nearby galaxies. Given the extreme physical conditions required if quasars really are at cosmological distances, this was not an unreasonable suggestion at the time, and it initiated an intense scientific dialogue that went on for decades. Finally, because there was no plausible physical basis for the proposition, because of the observational connections with more nearby active galactic nuclei, and because of convincing evidence for massive black holes in galactic nuclei, the idea has faded from serious consideration. I return to this point below.

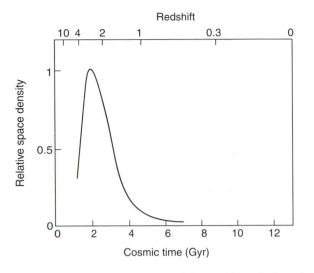

Figure 5.3 Schematic diagram of the cosmic evolution of quasars. The upper axis shows the redshift and the lower axis the age of the Universe. The number density of quasars appears to peak at a redshift of approximately 3 or at a cosmic age of 2 billion years. (From Shaver et al. 1996.)

More quasars were identified and redshifts measured. Rather quickly a global statistical aspect of these objects became apparent, largely through the work of Schmidt. The number density appeared to vary with cosmic epoch; there were more such objects at larger redshift, which is to say, in the past than at present. This is illustrated schematically in Figure 5.3 from a later publication (Shaver et al. 1996), where we see that the co-moving number density of quasars (i.e., correcting for the expansion of the Universe) appears to peak at a redshift of about 3 or when the Universe was only about 2 billion years old (about one-sixth of its present age). There appears to be a "quasar epoch" in the evolution of the Universe, and this was an early argument against a Steady-State Cosmology (the model wherein the Universe expands but always had the same mean density and appearance due to the continuous creation of matter).

5.2   Active Galaxies: Radio Galaxies and Seyferts

There are several reasons to identify quasars with galaxies. Similar forms of violent activity are clearly identified with galaxies and specifically with galactic nuclei. In 1956, before the discovery of quasars, Geoffrey Burbidge, then at Caltech, emphasized that the radio source Virgo A, identified with the luminous elliptical galaxy M 87, has an optical jet emanating from the nucleus and that the

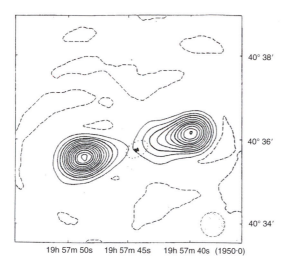

Figure 5.4 Early radio observations of Cygnus A at the upgraded 1-mile Cambridge interferometer by Ryle et al. in 1965. The contours show the lobes of radio emission, and the dotted ellipse between is the visible galaxy. The entire extent of the radio emitting lobes is enormous – 880 000 light years. This gives the impression of ejection from the galaxy by an impulsive event. Later higher resolution observations would give a different impression (Figure 8.2).

light from this jet is strongly polarized. On this basis he proposed that the emission mechanism, from optical to radio, is the synchrotron process, which means that the object is producing a very substantial energy in relativistic particles and magnetic fields (Shklovsky had earlier suggested the synchrotron process as the emission mechanism for the optical filaments in the Crab nebula, a supernova remnant).

In 1959 Burbidge extended this argument to other radio galaxies such as Cygnus A, associated with a galaxy of unusual morphology at a Hubble law distance of 600 million light years. As far as the radio morphology is concerned, this object resembles a number of quasars: it is a double radio source with the visible galaxy between the two radio emitting lobes (see Figure 5.4). At the redshift of Cygnus A, the lobes of radio emission are huge, extending to 150 000 light years on either side of the visible galaxy. Identifying the radio emission mechanism as synchrotron then implies, because of the enormous volume, that the total energy in magnetic fields and relativistic particles is at least 10^{60} ergs, a vast energy equivalent to the rest mass of one million suns (converted to energy by $E = Mc^2$).

Although radio galaxies typically do not exhibit the very high visual luminosity of quasars, the total energy in fields and particles is comparable to the energy emitted by a quasar over one million years; that is to say, the total energy requirements are no less severe than for the quasars, yet these sources are clearly identifiable with luminous massive elliptical galaxies that are not at extreme redshift. To Burbidge the morphology – jets and radio lobes beyond the visible galaxy – implied an eruptive or explosive phenomenon with impulsive ejection occurring often in two opposite directions, and the initial proposals were of single events of enormous energy – vast quantities of ionized gas, relativistic particles, and magnetic field ejected into the intergalactic medium by an extremely energetic explosion and confined by ram pressure due to the hypersonic motion through this ambient medium. Later higher resolution radio images would give quite a different impression, as discussed in Chapter 8.

But there is another sort of activity that is identified with the nuclei of galaxies such as the Milky Way – spiral galaxies. In 1943, a young astronomer, Carl Seyfert, working at Mt. Wilson Observatory published a paper on a class of spiral galaxies with optically bright nuclei exhibiting broad emission lines (see Figure 5.5). There were six galaxies in Seyfert's original list with hydrogen lines having widths up to 8500 km/s. These became known as Seyfert galaxies, and it is now known that several percent of spiral galaxies possess nuclei of this sort.

Figure 5.5 The Seyfert galaxy NGC 1068. Note the bright star-like nucleus. (Photographed at the Jacobus Kapteyn telescope. Courtesy of Isaac Newton Group of Telescopes and Nik Szymanek.)

In 1959 Lodewijk Woltjer, a former student of Oort who later emigrated to the United States, drew attention to Seyfert's generally ignored paper on emission line galaxies. Woltjer did not fit into the usual mould of Dutch astronomers at the time. His strength was not so much in planning and carrying out large observational and interpretive projects (although his dissertation on the Crab nebula was a definitive work), but in coming up with original and creative ideas of significant impact. He noted that the wide emission lines might be a result of gas motion in the gravitational field of a very massive and dense nucleus, and that such a mass concentration in the centers of galaxies might provide a source of high-energy phenomena. Following the discovery of QSOs it was realized that the nuclei of these objects shared many of the characteristics of QSOs, at least the radio-quiet variety. They are of course not so powerful, with luminosities extending up to that of an entire galaxy rather than 100 times a galaxy, but qualitatively, they appear to be quite similar with broad emission lines and a bright star-like nucleus. And this is activity clearly identified with otherwise quite normal low-redshift galaxies – galaxies like the Milky Way.

Then there was the case of the "exploding galaxy" M82, a nearby dusty disk galaxy seen edge-on with an extensive system of filaments extending above and below the plane of the galaxy. In 1963 Roger Lynds and Allan Sandage reported that the filaments were glowing in the light of the hydrogen H_α line and that Doppler shifts of this line implied that the filaments were moving away from the galaxy with a velocity on the order of 100 km/s. If this motion is along the rotation axis of the galaxy then it is almost perpendicular to the line of sight; the true velocity would be more like 1000 km/s, implying a very large kinetic energy of ejection (10^{55} ergs). It is now known that the gas filaments are being expelled in a much milder galactic wind being driven by a burst of star formation in the disk, but at the time, this object seemed to fit into the category of explosive events in galactic nuclei, the sort of events that presumably produce radio galaxies. In 1963, on the eve of the discovery of the large redshift of quasars, this evidence was complied by Geoffrey and Margaret Burbidge and Allan Sandage in a paper entitled "Violent events in the nuclei of galaxies."

5.3   Twinkle, Twinkle Little Quasar: Early Models of the QSO Phenomenon

A minimal energy requirement for quasars or radio galaxies is 10^{60} ergs or the equivalent of one million sun masses converted into energy with 100% efficiency. Stars like the Sun are converting mass into energy by the process of nuclear fusion in their hot dense interiors, but this process is only 0.7% efficient; that is to say, four hydrogen nuclei fusing into one helium nucleus loses only .007

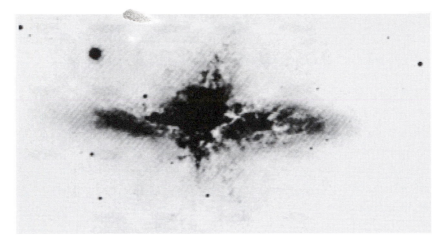

Figure 5.6 The "exploding galaxy" M82 in the light of the H$_\alpha$ line by Lynds and Sandage (1963). This gives the appearance of a galaxy shattered by a massive explosion. Reproduced by permission of the American Astronomical Society.

of its mass into energy. Therefore, to provide this total luminous energy emitted by a quasar would require more than 100 million solar masses of hydrogen being fused into helium. In ordinary stars like the Sun, this happens much too slowly, over a time scale of billions of years, so to produce the luminosity of quasars from ordinary stars, 10 to 100 galaxy luminosities would require at least 1000 billion stars, all within a radius of a few light months, the limit on the size of QSOs from the light variations. To support itself against gravity the stars in such a system would be moving at more than 90% of the speed of light and be so densely packed that every star would undergo a collision about once per year. If it took ten such collisions to totally disrupt a star then all stars in such a system would be converted into gas within ten years. Hydrogen fusion in ordinary stars is clearly an impossible mechanism for generating the high luminosity of quasars.

Then what about a single super-massive star of 100 million to one billion solar masses? Extremely luminous supermassive stars would be supported against gravity primarily by radiation pressure, and such configurations are also known to be unstable to gravitational collapse; which is to say, it would rapidly collapse to a more condensed configuration – or explode. What about a single super-massive star supported against gravitation by rotation? Again the problem is stability. Since the classical work on the stability of rotating fluid ellipsoids by Maclaurin and Jacobi it has been known that a substantial fraction of the support against gravity must be provided by thermal motion and not rotation. This brings us back to radiation-supported objects, which are known to be unstable. Instability was not objectionable to some (like Burbidge or Fred Hoyle and William

Fowler) who thought that radio galaxies might result from explosive or impulsive events.

All of these ideas were discussed in the early days of QSOs; instabilities and short-lived objects were acceptable for those who believed in explosive or impulsive events but could be problematic for those indications of longer term activity, such as Seyfert nuclei. As Woltjer pointed out, given that two or three percent of spiral galaxies have Seyfert nuclei, then the total lifetime of Seyfert activity must be at least several hundred million years (assuming that every spiral evolves through a Seyfert phase during its age of 10 billion years).

If collisions between stars occur in a dense stellar system, then why not use collisions to provide the energy source for QSOs and Seyferts? In 1964 in a letter to the journal *Nature*, Woltjer proposed that with densities on the order of 100 million stars in a region of a few light years, collisions would be sufficiently energetic to tear matter from stars. The cooling gas could possibly provide the high energies required for radio galaxies and quasars.

An isolated system of stars is expected to eventually reach a dense state because the stellar system dynamically evolves by two body gravitational encounters between stars (by gravitational encounter I don't mean an actual collision, but a very slight deviation of the star's orbit due to a perturbation by a distant passing star). This process, discussed separately by S. Chandrasekhar and Lyman Spitzer in 1940, causes the velocity distribution of the stars to evolve toward a Maxwell–Boltzmann form (like gas molecules in a closed jar) on a timescale, the relaxation time, that is proportional to the third power of the spread in stellar velocities (the velocity dispersion) and inversely proportional to the density; that is to say, the lower the velocity and the higher the density, the faster the system relaxes to this most probable state. During these gravitational encounters a small fraction of the stars attain a velocity that is higher than the escape velocity from the system, so during every relaxation timescale a few percent of the stars will escape. These escaping stars are near the top of the potential well, which means that this process removes mass but not binding energy: the same binding energy is shared by fewer stars, which implies that the system contracts, the density becomes higher, and the velocity dispersion becomes larger. The system digs itself deeper into the potential well and dynamically evolves toward a singular state as shown in Figure 5.7.

Thus, even an isolated stellar system with no dissipational gas dynamical processes (energy loss through shocks and radiation) will collapse to a singularity on some (generally very long) timescale. But of course, stars are not point masses; they are gaseous objects with extent and structure, and before the a stellar system contracts to the singular state, actual collisions, not just gravitational encounters, will become important. This was the basis of Woltjer's proposal.

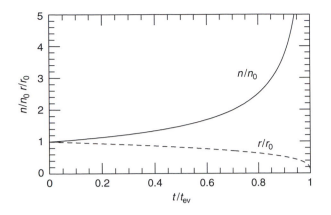

Figure 5.7 The time evolution of stellar density (solid curve) and radius (dashed curve) of a dense galactic nucleus. The density and radius are given in terms of their initial values, which may be on the order of 10 million stars in a region of one-third of a light year. The evolution time would be on the order of two billion years.

In 1966 the idea was taken up by Lyman Spitzer and William Saslaw, who constructed a more detailed model. Spitzer is well known for his fundamental work on plasma physics and the interstellar medium as well as his later work on the dynamical evolution of globular clusters. It is less well known that, in the mid-1960s, he was thinking hard about the source of high luminosity in QSOs, which was perceived to be a fundamental problem at the time. Spitzer and Saslaw (then an undergraduate student) devised a simple physical model for the mass loss that should occur during a collision between two stars and followed the evolution of a system so dense (near the end point of the evolution shown in Figure 5.7) that frequent collisions are occurring. They presumed that the gas liberated in such collisions will cool, collapse into a disk at the center of the system, and form new stars that would diffuse out into the system again. The radiation from the cooling gas and from the hot disk at the center of the system would be the source of the high luminosity of QSOs. Spitzer and Saslaw were rather unclear about the final state of such a system, but they did estimate that such processes could liberate up to 10^{12} solar luminosities (ten times the power of the entire Milky Way), comparable to that observed in luminous quasars. One problem, of which they were well aware, was the timescale involved. The initial system already must be in a very dense state – 100 million stars in a region of two or three light years – if this evolution were to occur in less time than the age of the Universe.

A different angle to this story was added by Stirling Colgate in 1967. Colgate was an expert on explosions of all sorts, and he pointed out that before a dense stellar system reaches the stage where stellar collisions are frequent but disrupting, it will evolve through a stage in which collisions are frequent but soft, with a collision energy less than the binding energy of the two stars, and this results in coalescence rather than disruption. The coalescence of colliding stars leads to the buildup of massive stars but, Colgate claimed, further coalescence of the most massive stars ceases when they reach a limit of approximately 50 times that of the Sun. Colgate argued that this end-point of mass buildup is due to the fact that the massive stars are bloated from the collisions; the coalesced objects contain the orbital energy of the two stars as well as their internal energy. An ordinary star then colliding with the bloated star passes directly through and no further coalescence occurs (like a bullet shot through a ball of cotton candy). But the resulting top-heavy mass distribution will lead to a greatly enhanced rate of supernovae. Supernovae are the most violent events observed to be occurring in stars; they are exploding massive stars that can liberate a substantial fraction of their rest-mass energy in a period a few weeks. If this process of buildup of massive stars actually occurs, then the supernovae rate from a dense galactic nucleus could be 5 to 10 per year, whereas in a normal galaxy like the Milky Way it is more like one or two per century. The luminosity of such a nucleus could easily achieve QSO values, and, what's more, a substantial fraction of the supernovae energy appears as synchrotron radiation, the form in which it is actually observed. This then appeared to be a very plausible model for the high luminosity of QSOs involving only known objects and conventional physics.

Would coalescence of colliding stars dominate over disruption? This was the question that I considered in 1969. Using a somewhat extended Spitzer–Saslaw model of stellar collisions along with a simple criterion for coalescence of colliding stars, I numerically simulated a dense stellar system undergoing frequent collisions. I found that for a certain range or properties of the stellar system coalescence does indeed dominate of disruptions but that Colgate had overestimated the effect of the bloating of coalesced stars – that in fact coalescence does not cease at 50 solar masses but continues at an accelerating rate because of the increase in geometric and gravitational cross section. There appears to be a coalescence runaway in a dense system of stars; the collision timescale becomes shorter than the stellar evolution timescale and there is no mechanism to stop the increasing stellar mass (see Figure 5.8). The implication is that stellar collisions will lead to a single massive star. Because of the instability of massive stars, it is also a mechanism for producing a massive collapsed object or an explosive event or both.

Martin Rees of Cambridge University has pointed out that the build up and subsequent collapse of massive stars through coalescing collisions is only one mechanism for the formation of a collapsed object in a galactic nucleus; the point is that in a dense system of the sort implied by observations of quasars it is difficult to avoid gravitational collapse to within a Schwarzschild radius. The possibility of

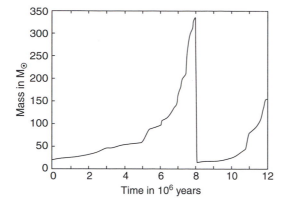

Figure 5.8 Development of maximum stellar mass in a system undergoing coalescing collisions. At a mass over 300 solar masses the star is arbitrarily destroyed and but the buildup begins immediately. This illustrates the phenomenon of runaway coalescence in a dense stellar system. (From Sanders 1970.)

quasars being such objects had been mentioned by Greenstein and Schmidt, but they proposed no mechanism by which such an object could radiate. In 1964, Edwin Salpeter of Cornell University, and, independently, the Soviet astrophysicist Yakov Zeldovich, demonstrated that a massive collapsed object in a galaxy such as the Milky Way would accrete gas from the interstellar medium. The gas would cool and collapse to the Schwarzschild radius and could emit a substantial fraction of its rest mass – between 5 and 20 percent – as radiation before it disappears beyond the horizon. This is at least ten times more efficient that nuclear fusion. Although Salpeter did not envision accretion through a disk (but rather the directed flow of a fluid about the gravitational field of such a collapsed object), he did show that it was possible for an object within its Schwarzschild radius – supposedly a dark object – to be in fact a source of high luminosity because of the extreme depth of its gravitational potential well.

5.4   Where Have All the Quasars Gone?

Terminology plays an extremely important role in the dissemination, popularization, and final acceptance of scientific ideas. This is particularly true of the term "black hole," first used by John Wheeler at a conference in 1967. Not only is it a handy abbreviation for "a collapsed object lying within its Schwarzschild radius," but it is also laden with imaginative, literary, and fanciful associations. By 1970 the expression was in general use even though black holes may be anything but black and may even be associated with the brightest objects in the Universe.

I have described in Chapter 1 the necessity of the black hole state as an end-point for the evolution of massive stars. Stars with mass greater than 1.4 solar masses cannot become white dwarfs, the quiet end-point of normal low mass starts, unless they lose substantial mass. This mass limit is about the same for neutron stars, the remnant for some stars that explode in a supernova. Because there are certainly stars of mass greater than this limit, then the expectation is that the Galaxy should be populated by stellar mass black holes. The observational consequences of such objects and their discovery is another story, but the point is that such bizarre objects, a consequence of Schwarzschild's solution to Einstein's equations, must exist.

But in the late 1960s there was still a reluctance among astronomers to accept this unconventional and, at that point, hypothetical object as an energy source for quasars, radio galaxies, and Seyfert galaxies. Hence the plethora of models involving supermassive stars, dense nuclei of stars with frequent collisions and/or multiple supernovae, and rotating massive disks, even though all of these configurations collapse on short order to the black hole state. Bizarre though it is, once formed, the massive black hole is stable and permanent. In fact, this very stability presented a perceived problem: From Figure 5.3 we see that there were far more quasars in the past than at present. Because black holes do not disappear, where are all of those objects that were so brightly shining 10 billion years ago? Where have all the quasars gone?

An answer to this question was provided in 1969 in a major contribution by Donald Lynden-Bell (Figure 5.9), then at the Royal Greenwich Observatory. Lynden-Bell's remarkably simple answer was that the old quasars are still with us (like dinosaurs in the form of birds); the black holes are present in the nuclei of ordinary galaxies; they are just not accreting gas at the high rate of the past.

Salpeter had demonstrated that taking an element of gas from a large distance down to several Schwarzschild radii can generate heat by means of friction with itself and, because of this heat, radiation. It is possible to radiate up to 20% of the rest mass energy of the element of gas and the source of this energy is gravitational field of the black hole. This is the most efficient means of converting mass into energy, apart from total annihilation of matter and antimatter. Moreover it was directly observed that the source of all of this radiation is a very small region, less than a light year and probably on the order of the size of the Solar System. To produce the equivalent of at least one million solar masses of energy (by $E = Mc^2$) at 10% efficiency would require at least 10 million solar masses, all in an extremely compact region. It is difficult now to imagine now that black holes were not immediately seized upon as the obvious explanation, but we should recall that at the time black holes were not entirely reputable objects (some astronomers of the

Figure 5.9 Donald Lynden-Bell as he appeared about the time of his proposal on the presence of black holes in normal galactic nuclei. Not only is he a leading expert in mathematical methods of physics, but he is also a highly creative scientist with a contagious enthusiasm. (Courtesy of Donald Lynden-Bell and Cambridge University.)

time were undoubtedly not excited but rather put off by this fanciful terminology; in fact Lynden-Bell did not use the expression).

Lynden-Bell proposed that most galactic nuclei have at least a small old quasar at their centers. In the past, when galaxies were being formed, there was much more gas to be consumed. When the Universe was two to three billion years old, this was the epoch of rapid accretion and growth of black holes – the quasar epoch. These black holes are still present, but underfed. The black hole is the engine, but the engine requires fuel to work, and there is just less fuel available now than in the past. At present, the fueling of black holes in a galaxy like the Milky Way is the slow accretion through a disk, at a rate much less than one solar mass per year. After all, the Galaxy has a moderate nonthermal radio source at its center, Sgr A, and the limit on a massive object at the center from the 21-cm line rotation curve of Rougoor and Oort was about 10 to 100 million solar masses. So it is reasonable to assume that there is a black hole of this mass in the Galactic nucleus. Lynden-Bell predicted that there should be dynamical evidence for a mass concentration of this order at the center of the Galaxy, and much of the remainder of this story concerns the quest for that evidence.

5.5   Summing Up a Decade of Discovery

Astronomy is a science generally driven by observations rather than by theory. New technology such as radio astronomy leads to the discovery of new phenomena such as quasars and opens intense discussion over alternative models. With respect to massive black holes, who could have possibly predicted the existence of such objects even though the theoretical framework had been in place for fifty years? It took the crisis provoked by the implied extreme properties of quasars – the power of 100 galaxies emerging from a region less than one light year – to provide the impetus for invoking these bizarre, previously purely theoretical, constructs.

When such a crisis emerges, developments may proceed rapidly. Ten years passed between the publication of the 3C catalogue of discrete radio sources and the publication of what would become the definitive model of activity in galactic nuclei. Milestones along this decade long journey were the optical identifications of several radio sources with stellar like objects, others with elliptical galaxies; the discovery of the nature of quasar spectra as highly redshifted hydrogen lines; the realization of the similarity of these objects to emission line galactic nuclei, Seyfert galaxies, in the nearby Universe; the emergence of alternative models involving supermassive stars, disrupting stellar collisions in dense systems, coalescing collisions, and multiple supernovae and the realization that all of these tend toward gravitational collapse on short order; the realization that black holes are not black at all if in an environment where gas can be rapidly accreted; and the emergence of the idea that inactive massive black holes may be present in most galactic nuclei including that of the Milky Way.

This was certainly not a simple straight-line development. One can trace two major lines of thinking on these new classes of objects: explosive or impulsive phenomena ejecting ionized gas, magnetic fields, and relativistic particles on short timescales and favoring models involving unstable objects; and longer term, more steady processes such as accretion onto black holes, although possibly episodic, as well as outflow through continuous jets rather than explosive ejections. Although subsequent observations would single out the second interpretation, the 1960s was a period of intense conflict between these ideas.

In general, the fundamental perception of the Universe and its visible contents changed in this decade. Previously, the evolution of galaxies, unlike that of stars in some cases ending in violent supernova explosions, was thought to be rather passive since formation, proceeding over billions of years without much display of spectacular activity. But by 1970, galaxies, at least their nuclei, were known to exhibit a range of activity: ejections, explosive or continuous – radiant energy exceeding that of an entire galaxy and emerging from a compact region – Seyfert or quasar phenomena operating on timescales as short as one million years. The Universe of galaxies came to be perceived as quite violent.

The story was certainly not over and all questions settled by 1970. The massive black hole model was by no means accepted. Arguments over alternative models such as massive rotating objects or gas disks continued into the 1980s and beyond, as did heated discussions over explanations based on new but unspecified physics. The relationship among the various manifestations of active galactic nuclei, QSOs, radio galaxies, Seyfert galaxies, was still to be understood, as well as the connection of star formation with active nuclei. And observations of the central region of the Galaxy would come to play a significant role in the clarification and resolution of these issues. But first the veil covering the inner light years of the Milky Way had to be parted further.

6

New Windows on the Galactic Center

6.1 The Near Infrared and the Distribution of Stars in the Galactic Center

At 26 000 light years the center of the Milky Way Galaxy is the closest galactic nucleus. At that distance one light year corresponds to about 8 arc seconds, which is easily resolvable with large optical telescopes; in fact, such instruments can resolve one arc second, which would correspond to one and one-half light months at the distance to the Galactic Center. Even the Schwarzschild diameter of a four-million solar mass black hole would have an angular size of 20 micro-arc seconds (20 millionths of an arc second), which can be resolvable by VLBI (very long baseline interferometry) at a wavelength of 1 millimeter. So if one wishes to observe a galactic nucleus in fine detail, the center of the Milky Way is the place to look.

This was realized by Jan Oort, who in the 1970s became intensely interested in the Galactic Center and the processes that might be going on there. Oort was impressed by the number of gas features observed in the 21-cm line of neutral hydrogen in the general direction of the center that appeared to be expanding away from the center – features such as the 3-kiloparsec arm, seen in absorption at −53 km/s (motion toward the Sun and away from the center), and a less prominent feature apparently on the opposite side of the center but moving away at 135 km/s. For Oort, this was the closest example of ejections or expulsions from the region of a galactic nucleus – the sort of phenomena thought to be observed in more distant active nuclei. But Oort also realized that the largest optical telescope could not penetrate through the obscuring dust between the Sun and the Galactic Center, which dimmed the visible light by a factor of 10^{-12}. That is to say, only one in one trillion photons can penetrate this veil.

In the infrared, however, at wavelengths of 1 to 2 micrometers, this dimming is only a factor of ten – that is, 10% of the radiation emitted at the center arrives at Earth. This is because the small particles in the interstellar medium, with sizes comparable to or smaller than the wavelength of light, are very effective in scattering and absorbing radiation of short wavelength (light or ultraviolet) but much less effective in blocking longer wavelength radiation, radio or infrared.

All objects with finite temperature emit continuum electromagnetic radiation – the cooler the object, the longer the wavelength of the typical radiation. The Sun, for example, with a temperature of 5800 K, emits continuum radiation peaking at about 0.5 micrometers, or "microns," which our eyes detect as yellow light (one micron is one millionth of a meter). Our bodies, at a temperature of around 300 K emit radiation around 10 microns, that is, in the infrared, which our eyes cannot detect. Many astronomical objects, such as cool stars and hot dust, emit radiation primarily in the infrared. In particular, giant stars with temperatures between 2500 and 4000 K are emitting much of their radiation between 1 and 2 microns, the *near* infrared. These are the sorts of stars that might be expected to contribute significantly to the radiation from an old population of stars like those in the Galactic bulge. So in order to trace the distribution of typical stars in the Galactic Nucleus, the near-infrared is the ideal wavelength range for observation. The problem is that the traditional astronomical detectors, the human eye and photographic plates, are insensitive to this radiation. How do we detect infrared radiation?

Infrared astronomy emerged in the 1960s because infrared detectors became readily available. This again was because of technology prompted by war or threat of war. The detection of warm human bodies at night is of obvious military significance, as is the construction of missiles that can detect and follow the heat generated by aircraft jet engines.

The detection of infrared radiation was greatly facilitated by the development of semiconductors. In a cooled semiconducting material such as lead sulfide, the electrical conductivity is extremely sensitive to the flux of infrared radiation between 1 and 3 microns – the more infrared photons, the higher the conductivity. Small variations in the infrared flux on a lead sulfide chip are easily detected by passing a current through the semiconductor.

By the mid-1960s this technology had advanced and been declassified sufficiently that it could be taken over and applied by astronomers, and observations of astronomical objects became possible – possible but not easy. It is somewhat ironic that after traveling relatively unobscured all the way from the Galactic Center, the last 20 kilometers of Earth's atmosphere becomes the essential problem for the passage of infrared radiation. The atmosphere is generally opaque to the infrared because of absorption by molecules, primarily water vapor. So to detect this extraterrestrial radiation, the astronomer needs to observe from a location that is high and dry – usually not a problem because most major observatories are located in such environments. But obviously flat damp Holland, at its northern latitude, is an inappropriate site for infrared observations. This, and the fact that the detector technology had been developed primarily in the United States, meant that Oort could not directly exploit this particular opening through the veil of interstellar dust. The center of activity moved to the United States.

There are certain wavelength windows in the atmosphere where this absorption is less problematic. The water vapor molecules absorb in definite wavelength bands – much broader in frequency or wavelength than atomic spectral lines; the observer should avoid these bands. It is fortunate that one of these windows between the molecular bands lies in the near-infrared and another in the mid-infrared range (around 10 microns). For longer wavelengths, the far-infrared around 100 microns, the solution is to move above much of the atmosphere and observe from high platforms such as balloons or satellites.

The very first near-infrared observations of the Galactic Center (at 2.2 microns) were those of Eric Becklin and Gerry Neugebauer carried out at Mt. Wilson and Palomar Observatories in California. These observations revealed an extended source centered on the radio source Sgr A and elongated with respect to the Galactic plane. With reasonable assumptions about the extinction law (how the extinction by dust varies with wavelength) and the intrinsic spectrum of the source (similar to the starlight from the bulge of the Andromeda Galaxy), this observed source is consistent with being emission from the old population of stars in the central bulge of the Milky Way out to a radius of 100 light years.

The implied mass density distribution in this region is of power law form, that is,

$$\rho = \rho_0 (r_0/r)^{1.8} ,$$

which is to say, the mass density in old stars declines from a central core with density of ρ_o and a radius of r_0 almost as the inverse square of distance from the center. The constants in this relationship, ρ_0 and r_0, are somewhat uncertain owing to the unknown mass-to-light ratio of the stellar population in the center and the precise distance to the center, but taking the present assumed distance to the center of 26 000 light years and a mass-to-light ratio of 1 to 2 (in solar units), then we find $\rho_o \approx 50\,000$ solar masses per cubic light year and $r_0 \approx 1$ light year. At 1 light year from the Galactic Center the stellar density is three million times larger than the density of stars locally (near the Sun). The total mass within 1 light year would be 200 000 and 500 000 solar masses depending on the actual value of r_0 – whether or not the same density law extends right into the center. This means that the average distance between stars would be one-hundredth of a light year (in

Figure 6.1 The near-infrared view of the Galactic Center. This is a more recent image made by Blum, Ramirez, and Sellgren using the Ohio State Infrared Imager/Spectrometer (OSIRIS) and shows the distribution of individual bright stars. The scale of the figure is 1.7 by 1.7 degrees, corresponding to about 800 light years on one edge at the distance to the Galactic Center. In this, and following images of the Galactic Center, the galactic plane runs diagonally across the figure from upper left to lower right.

the neighborhood of the Sun the average distance between stars is about 4 light years).

This does appear to be a comparatively extreme environment as far as the density of stars is concerned. On the other hand, the rate of collisions between stars in the inner light year would be less than one per 1000 years, so it is quite unlikely that this process would have importance in generating energy or changing the mass distribution of stars. On an Earth-like planet orbiting a star in the central light year, the sky would be full of extremely bright stars; the total starlight falling on the planet from ordinary low-mass stars would exceed that of the full moon by a factor of 20. Still, most stars would appear to be unresolved pinpoints of light but with rather large proper motions – changing their positions in the sky by more than one minute of arc every year.

It is, however, improbable that planetary systems like the Solar System could exist in this inner light year of the Galaxy. A passing star would come within the orbit of the Earth once every 10 million years and it would probably not take more than ten such passages to eject the planet from the orbit about its star; certainly life could not evolve in such an changeable environment.

If this same density law continues out to distance of 2000 light years, and there is more recent evidence that it does, then the total mass within this radius would be five billion solar masses. This would yield a rotation velocity of 185 km/s at this radius, considerably below that estimated by Rougoor and Oort on the basis of the

21-cm line observations but consistent with the measured rotation velocity further out at 10 000 light years. The Rougoor-Oort measurement assumes that the gas is moving on circular orbits, and, as discussed in the following chapter, this is most likely not the case. The issue was the source of some controversy for twenty years.

The observations of Becklin and Neugebauer were the first revelation of the distribution and high density of stars in the inner light years of the Milky Way; in fact, subsequent near-infrared observations by a host of instruments have essentially confirmed the form of the density law, although there has been continued discussion over its normalization, ρ_0 and r_0. But is there evidence in these observations for anything more bizarre – a black hole, for example? Something that might be the source of all of these ejections that Oort was concerned about. Becklin and Neugebauer detected a bright point source near the center of the extended infrared emission, but its colors were consistent with being a luminous giant star near the Galactic center. For evidence of an unusual object we move again to the radio continuum and specifically the technique of interferometry.

6.2   A Unique Source at a Unique Location

With great foresight and drive Jan Oort had pushed the Netherlands into radio interferometry with the planning and construction of the Westerbork Radio Synthesis Telescope, WRST (located in Drente south of the "hundsrug"), a linear array of fourteen dishes of 25-meter diameter with a maximum east–west baseline of 1.5 km. Coming online in 1970, the WRST was ten years ahead of the planned Very Large Array (VLA) in the United States, and, although the VLA was to be much larger and more powerful, WRST was first and scooped off a number of major discoveries before the VLA (flat rotation curves of spiral galaxies, jets in extragalactic radio sources).

In hiring astronomers, the Netherlands has never been particularly nationalistic and so, when WRST came online the Universities of Groningen and Leiden reached out around the world to hire experts in radio interferometry. One of these was Ron Ekers, an Australian who had been a postdoctoral fellow working at the Owens Valley interferometer of Caltech. There he met Donald Lynden-Bell, then on sabbatical leave, who encouraged Ekers to look for a compact radio source at the Galactic Center, the sort of signal that might be associated with a black hole.

Sagittarius A (Sgr A) is the source of radio continuum radiation at the Galactic Center and, with single dish radio telescopes, is seen to be extended along the Galactic plane, with an angular size of 3 to 4 arc minutes (22 to 30 light years at the distance of the Galactic Center). The peak of the intensity of radio emission corresponded to the peak of the Becklin-Neugebauer near-infrared source – the peak of the starlight. So Ekers used the Owens Valley interferometer to look for a more compact radio source at this position. He found an unresolved source, but the interferometer beam pattern (the shape of the region resolved by the interferometer) was too complicated to say much about its structure. The problem was that the Owens Valley interferometer had only two elements, only two telescope dishes. Although high resolution is possible with two elements, it is difficult to get a true picture of the distribution of radio emission. It would be as though one were observing with an optical telescope, a reflector for example, with masking tape over 90% of the surface; the picture would be blurry and complicated. Given this incomplete picture of the radio sky, Ekers and Lynden-Bell concluded that the small radio source in Sgr A was consistent with being a compact region of ionized hydrogen, perhaps around a newly formed star, such as seen in other parts of the Galaxy. There was, they concluded, no evidence here for a massive black hole.

Then when Ekers moved to Groningen in 1972, he decided to apply the WRST to observing the radio source at the Galactic Center. With fourteen dishes Westerbork was a very complete array; going back to the analogy with optical telescopes it would be as though only 30% of the aperture were covered with tape so it is possible to get a more complete picture at radio wavelengths. The problem with WRST was that the very northern latitude of Westerbork and the location of Sagittarius in the very southern sky gave a interferometer beam that was extremely elongated north–south. There was essentially no resolution in this north–south direction. Ekers and his collaborators solved this problem by combining the Westerbork observations with those at Owens Valley. Westerbork had the more complete coverage and Owens Valley had the more southern latitude; the deficiencies of each system were made up for by the other and with this combined system they mapped the structure of Sgr A in detail.

Sgr A turned out to be a complicated multiple source. Broadly speaking it separates into two components, Sgr A East and Sgr A West. "East" is a nonthermal source (synchrotron radiation), and displaced from the peak of the near-infrared emission by about 2 minutes of arc, a projected distance of 15 light years. It is probably a background supernova remnant and not directly connected to Galactic Center phenomenology. Sgr West is a thermal source (continuum radiation from hot gas) and peaks at the position of the infrared center, and so arises at the very center of the Galaxy. These two components are shown in Figure 6.2 taken from a later publication of Ekers and collaborators. The intense structure on the right is Sgr A West, the thermal emission from hot gas, and the shell-like structure on the left is the nonthermal emission from the Sgr A East. But still there was no evidence for a compact, unresolved radio source that might be associated with a black hole.

In 1974 Bruce Balick and Robert Brown of National Radio Astronomy Observatory (NRAO) looked more closely at Sgr W using an upgraded radio interferometer

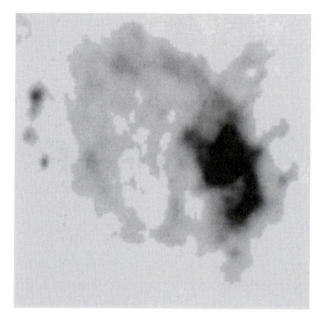

Figure 6.2 The complex nature of Sgr A is shown in this radio map (20 cm) by Ekers et al. 1983. Top to bottom corresponds to about 30 light years at the distance of the Galactic Center. The intense source on the right (Sgr A West) is thermal emission from ionized gas (shown in more detail in Figure 6.5). The shell structure on the left is nonthermal emission from a proposed nearby supernova remnant, Sgr A East.

at Greenbank, West Virginia. The Greenbank interferometer consisted of three 26-meter dishes with a maximum baseline of 2.7 kilometers and a new 14-meter telescope located about 35 kilometers southwest on a mountain top. The more dense array provided a reasonably complete coverage of the sky and the more distant telescope provided higher resolution. With this system Balick and Brown discovered a source of intense emission at a wavelength of 3.8 and 11 cm. This source was smaller than one-tenth of an arc second or about one-hundredth of a light year at the distance of the Center. The total power of at radio wavelengths was roughly equal to ten suns; although this is much lower than the radio power of active galactic nuclei, it is more powerful than any other compact radio source in the Galaxy. For example, the most luminous pulsars, the dense spinning neutron stars that are remnants of supernovae explosions, have a radio power that is 10 000 times less than that of the compact Galactic Center source. It is, in the words of Martin Rees, a unique source in a unique location.

Although it was not generally appreciated at the time, Balick and Brown had obtained the first direct glance at the Galactic Center black hole. The compact radio source quickly became designated Sgr A*, pronounced "Sag A star." Its location was consistent with being near the peak of the extended near-infrared source – the distribution of bulge stars – but did not coincide with the original point source of Becklin and Neugebauer.

In 1979 Donald Backer and Richard Sramek of the NRAO began a program to measure the proper motion of Sgr A* using the Greenbank interferometer. This is an important observation because if Sgr A* really is associated with a massive object, then it should be sitting still in the middle of the potential well of the Galaxy and not jittering about like a solar mass pulsar. When they analyzed their results they discovered that it was moving with a velocity between 200 and 300 km/s, but in the plane of the Galaxy and opposite the direction of the Sun's revolution about the center. They had directly observed the rotation of the Galaxy at the position of the Sun! The peculiar motion of Sgr A* itself was consistent with being zero (albeit with large errors).

Subsequent observations by Mark Reid and collaborators using the technique of very long baseline interferometry (VLBI) confirmed this result with high precision. If the Sun is revolving about the Galactic Center, and the distance to the center is 26000 light years, then the circular velocity at the position of the Sun is 220 km/s, in quite nice agreement with the original Dwingeloo measurement of the rotation curve of the Galaxy (Figure 4.6). It is amazing that this rotation motion of the Galaxy could be so directly observed by looking at the point source in the Center. It also presents strong evidence that the object associated with Sgr A* is more massive than 1000 solar masses.

6.3   Mid- and Far-Infrared Radiation

At mid-infrared wavelengths, around 10 microns, the emission is no longer dominated by direct heat radiation from stars but from dust in the vicinity of hot stars – dust heated by starlight up to temperatures of approximately 300 K. In 1973, George Rieke and Frank Low (Low was one of the early pioneers of infrared detection technology) at the University of Arizona produced the first detailed map of the inner 2 light years of the galaxy in this mid-infrared region. They found a number of discrete sources lying along an extended ridge of emission. A year later this was followed up by Becklin and Neugebauer at Palomar in observations at 2.2 and 10 microns; they found that the infrared emission within 6 light years of the center was actually dominated by point sources, not the smooth extended component from the old stars (see Figure 6.3, which is the distribution of stars in the central light years as observed 20 years later by Genzel et al.). They identified nineteen distinct sources (designated IRS for "infrared source"), the brightest of which at 2.2 microns was the point source in their original observations, IRS 7.

Figure 6.3 A more recent near-infrared view of the surroundings of Sgr A* (from Genzel et al. 2003). Here it is evident that most of the emission from this stellar core is from individual bright stars. The bright object in the upper center is IRS 7, the giant star originally observed by Becklin and Neugebauer, and the arrows indicate the position of Sgr A*. The small cluster of stars nearby are IRS 16.

Their map at 10 microns is reproduced in Figure 6.4, where we see the north–south ridge of sources; at 10 microns, IRS 7 is no longer the brightest. The compact radio source Sgr A* does not coincide with any of these strong point-like near infrared sources but is roughly at the position of IRS 16, 5 arc seconds south (about one-half light year) of IRS 7 (denoted by the X in 6.3). Becklin and Neugebauer concluded that these mid-infrared discrete sources are probably arising from dust surrounding luminous stars; the ridge of extended emission is a more general dust distribution mixed with ionized gas. This ionized gas-dust ridge appears to be wrapping around Sgr A – a highly suggestive morphology.

In order to determine the mass distribution it is necessary to measure velocities. In this respect, Nature has generously provided a spectral line in the mid-infrared – the 12.8-micron line of singly ionized neon (neon with one electron removed). This line was first observed in 1976 by a group at Lick Observatory in California – a group (Wollman et al. 1976) established by Charles Townes, who was known for his development of the "maser" (more on natural masers in Chapter 9). The neon emission line was mapped along the north–south ridge in Figure 6.4.

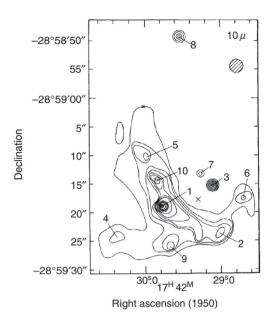

Figure 6.4 The 10-micron map of the inner light years of the Galactic Center, by Beklin and Neugebauer (1974). The position of Sgr A* is marked by an X. Note that this does not correspond to any bright 10-micron source; these are due to dust heated by bright stars. The distance along the north–south ridge, from IRS 5 to IRS 2, is about 20 arc seconds corresponding to 2.5 light years. The figure is taken from Oort 1977, who added the position of Sgr A*.

Along this feature, a distance of about 2.5 light years, the velocity appeared to vary systematically from −200 km/s to +100 km/s. If interpreted in terms of orbital motion in a gravitational field this would indicate a mass of a few million solar masses within half a light year – the first actual indication of a mass concentration at the Galactic Center. The emission, however, does arise in gas and gas is subject to forces other than gravitational – radiation pressure, hydrodynamical disturbances – so this was not yet definitive. The definitive determination would come with the observations of stellar motions.

In the early 1970s the far infrared emission from the Galactic Center region was observed from balloon-borne telescopes and detectors consisting of cooled bolometers (the electrical conductivity depends on temperature, which is raised when exposed to infrared radiation). In 1971 the first 100 micron map of the central region was published by a group led by William Hoffmann at NASA. An extended source of far-infrared emission was detected over the inner 1600 light years. This emission is primarily from cool dust in the interstellar medium (100 to 200 degrees Kelvin), not dust in the near vicinity of bright stars like in the mid-infrared. The total luminosity of this source was consistent with the total luminosity of stars over this region – the far infrared is thermal re-radiation of starlight by dust particles. This observation gave added confidence to the density distribution of stars implied by the near-infrared radiation.

6.4   The "Mini-Spiral" in Sgr A

The center of the Milky Way is a remarkable region. Just when we think that we are understanding the Galactic Center phenomena, the observations throw up another surprise that complicates the view further. So was it with the radio continuum in the early 1980s.

In 1980 the Very Large Array (VLA) of the National Radio Astronomy Observatory was inaugurated in the New Mexico desert. This facility represented the prime investment of the United States in radio astronomy up to that time – an array of twenty-seven paraboloid dishes of 25-meter diameter each and capable of sub-arc second resolution at a wavelength of 6 centimeters. In 1980, the instrument was ready for use, and one of the first sources to be observed was, of course, the Galactic Center.

In 1981 Ron Ekers, by then director of the VLA, and Jacqueline van Gorkom, of the National Radio Astronomy Observatory (formerly of the Kapteyn Institute in Groningen), along with Ulrich Schwarz and Miller Goss of the Kapteyn Institute, observed the spatial distribution of continuum radiation from Sgr A at wavelengths of 6 and 20 centimeters. The structure they observed in Sgr A West, in the near vicinity of the compact source Sgr A*, is shown in Figure 6.5 (see also the color representation on the back cover of this book). The remarkable morphology appears to be that of a spiral, but a spiral less than 10 light years from north to south – a mini-spiral. The emission is from ionized gas, with a total mass estimated to be less than 50 times that of the sun.

Ekers and collaborators suggested several possibilities for this surprising morphology: a jet of material ejected into a medium rotating about the center and bent by this rotating ambient matter; precessing jets, plausible because other radio sources, galactic and extragalactic, show evidence for precessing beams; tidal distortions of infalling small clouds, also conceivable because there are a number of clouds of all sizes in the central region exhibiting highly random motion. Looking back at the 10-micron observations (Figure 6.4), we can clearly identify the north–south feature in the radio continuum with that of the main ridge of mid-infrared radiation; even several of the individual sources (hot dust around bright stars) in the near infrared are evident in the radio continuum. As in the mid-infrared, the ionized gas as traced by the radio continuum appears

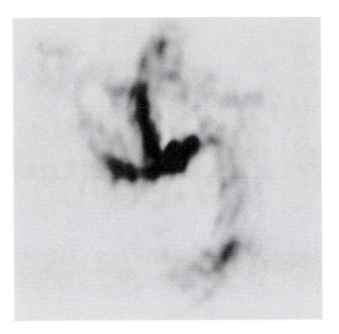

Figure 6.5 The radio map of Sgr A West made at the VLA by Ekers et al. (1983) at a wavelength of 2 cm. This shows the distribution of hot ionized gas in the inner 10 light years. This is shown on the rear cover in a color representation. In the color figure the bright spot is the compact radio source Sgr A*. This image was made by D.A. Roberts, W.M. Goss, and F. Yusef-Zadeh and is reproduced here by courtesy of NRAO/AUI.

to wrap around the point source Sgr A*; this supports the third alternative explanation of Ekers and collaborators – that we are seeing the ongoing accretion onto the central black hole. A greater understanding of the morphology of the ionized gas would develop over the next decade.

The relation of this mini-spiral to the overall Sgr A source (East and West) is evident in Figure 6.2.

6.5   Molecular Clouds in the Central Region

In the early 1970s it became evident that star formation in the Milky Way occurred in the environment of dense molecular clouds. These are clouds or concentrations of interstellar matter that is so dense that the neutral atoms combine on the surface of dust grains to form the hydrogen H_2 molecule. Normally H_2 is dissociated in the interstellar environment through the action of the background ultraviolet radiation from hot stars, but in the cool interiors of these dense clouds, the hydrogen molecules are shielded from this destructive radiation. Because of its symmetry the H_2 molecule is difficult to detect, but a

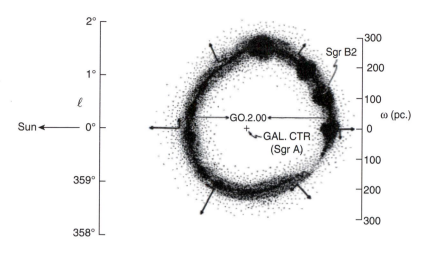

Figure 6.6 The Scoville (1972) model for the distribution and motion of molecular clouds in the inner few hundred light years of the Galaxy: the "expanding ring."

number of other molecules, such as formaldehyde (CH_2O) and carbon monoxide (CO), can be detected at radio wavelengths. CO is a particularly ubiquitous and robust molecule with a conspicuous spectral line at a wavelength of 2.6 mm. By 1970, radio receiver technology had advanced to the point that this line could be readily observed in the interstellar medium, arising primarily in molecular clouds. The CO, formaldehyde, and other molecular lines opened yet another window on the Galactic Center because the central several hundred light years of the Galaxy contain complexes of massive molecular clouds.

Looking at the 21-cm absorption line spectrum against Sgr A (Figure 4.8) we see that there are features at about 20 km/s and 50 km/s. These are actually associated with massive molecular clouds in the direction of but slightly displaced from Sgr A. Another molecular concentration, at a projected distance of 390 light years from the center, is Sagittarius B2, one of the largest molecular clouds in the Galaxy. The total mass of Sgr B2 is several million solar masses and the average density is 3000 hydrogen atoms per cubic centimeter. It is about 150 light years across and a chemical cauldron synthesizing molecules of all sorts – one-half of all molecular lines detected in the Galaxy have been discovered in this cloud. The presence of such massive and dense molecular clouds in the central region of the Galaxy suggests that star formation is actively occurring there.

One of the first systematic studies of the distribution of molecular clouds in the central Galaxy was provided by Nicolas Scoville in 1972 while he was a graduate student at Columbia University. He saw that the positions and kinematics of molecular clouds in this region appeared to be consistent with being distributed in an expanding rotating ring between 350 and 450 light years from the center. At the same time this structure was independently identified by Norio Kaifu, Takaiji Kato, and Tetsuo Iguchi of the University of Tokyo. The expansion velocity is around 100 km/s, and with a mass estimated to be on the order of three million solar masses, this gives a kinetic energy of expansion of close to 10^{54} ergs (the rest mass energy of a solar mass star). This is reminiscent of other expanding feature seen in neutral hydrogen (but further out) and suggestive of explosive events occurring in the Galactic Center, a favorite idea of Jan Oort but an ongoing controversy. If this feature has its origin in a symmetric explosion at the center of the Galaxy, the actual energy would be perhaps 100 times larger than the observed kinetic energy, so it would be a significant event.

We now turn to this issue of explosions and the remarkable fact that observations of the Galactic Center following up on this idea have Revealed a very basic aspect of the structure of the Milky Way Galaxy.

7

The Milky Way as a Barred Spiral Galaxy

7.1   Ejections and Explosions

The discipline of astronomy is replete with unconventional personalities. The Armenian astrophysicist Victor Ambartsumian (Figure 7.1), who had a highly respectable career behind him by the mid-twentieth century, was certainly one such character. In 1958 he presented a paper at the prestigious Solvay conference in Belgium on the role of activity in galactic nuclei in shaping the structure and evolution of the surrounding galaxy; he proposed that spiral structure is formed by ejections from the nuclei of galaxies and that, in extreme cases, new blue galaxies are "born" (ejected) from the centers of giant galaxies. This all seemed quite bizarre at the time but the concurrent discovery of radio galaxies and, several years later, of quasars gave support to the less radical idea that many galaxies appear to undergo sudden impulsive events involving mass ejection, often in two opposite directions.

This was an initial interpretation of the morphology of radio galaxies, where two enormous lobes of radio emission are often observed beyond and on opposite sides of the visible object, generally an elliptical galaxy (see Figure 5.4 of Cygnus A). An early model involved the explosive ejection of two oppositely directed clouds of ionized gas and relativistic particles that are confined by ram pressure of supersonic motion through an ambient intergalactic medium. The amount of energy involved in such events would be huge, much larger than the rest mass energy of one million suns directly observed as relativistic particles. It was these large energy and mass requirements that led some astronomers like Ambartsumian to consider the idea of totally new physical processes such as the creation of new matter in the central regions of galaxies like the Milky Way – processes without any basis in current physics.

Figure 7.1 Victor Ambartsumian, the Armenian astrophysicist who first emphasized the impact of an active galactic nucleus on the surrounding galaxy. (AIP Emilio Segre Visual Archives, Physics Today Collection.)

This idea of ejections appeared to be supported by more local examples such as the "exploding" galaxy M82, which was only ten million light years away (see Figure 5.6). Here, a large system of filaments shining in the light of the red Balmer alpha hydrogen line extended above and below the disk of the galaxy; the velocity of these filaments with respect to the disk of M82 was deprojected (incorrectly as it turned out) to be more than 1000 km/s, implying an event of enormous energy. And then, of course, there are the features toward the center of the Milky Way that appear to be moving away from the Galactic Center with velocities of 50 to 150 km/s. Was this another example of expulsive phenomena? Was this sort of phenomena general?

Jan Oort thought so. And he thought that our, relatively speaking, local neighborhood was the ideal environment to study such phenomena. In the late 1960s he asked his student Piet van der Kruit to look again at the central region of the Galaxy with this possibility in mind. Van der Kruit, using the Dwingeloo telescope, re-observed the central 20 degrees of the plane of the Galaxy in the 21-cm line of neutral hydrogen, but this time he also looked 5 degrees above and below the Galactic plane. He noticed that those features with velocities forbidden in the sense of Galactic rotation (implying noncircular motion) tended to lie out of the plane in two opposite quadrants; this he took as evidence for ejection in two opposite directions. To produce features such as the 3-kiloparsec arm (the feature at a radial distance of 10 000 light years from the center apparently moving outward with a velocity of 53 km/s), van der Kruit constructed a model involving the expulsion of clouds from the center having a total mass of about 10 million solar masses. These clouds are ejected at an angle of 25 to 30 degrees from the plane with velocities on the order of 600 km/s. The total kinetic energy in these clouds would then be that of the rest mass of 10 solar masses (10^{55} ergs), considerably less than that of a radio galaxy but 1000 times larger than that of an energetic supernova. When the clouds fall back into the plane and interact with the rotating interstellar medium they can reproduce features such as the 3-kiloparsec arm without disrupting the rotating material further in – the so-called nuclear disk.

Then, in 1972, van der Kruit, Oort, and Mathewson, using the recently completed Westerbork array, found another example of what they interpreted as ejection from the nucleus of a "normal" spiral galaxy, NGC 4258. The spiral arms present in many disk galaxies are luminous structures shining in the light of newly formed stars. The arms may be tightly wound, which is to say, almost circular, or loosely wound – very open in appearance. NGC 4258, shown in Figure 7.2, appears to be of the second sort; at visible wavelengths two open arms delineated by bright stars and regions of ionized hydrogen wind from the central regions of this galaxy. But, as observed by van der Kruit and collaborators, in the radio continuum at a wavelength of 20 cm, an additional pair of arms appears to emerge from the nucleus – a set of spiral arms that had been previously detected in faint smooth emission of the red hydrogen line and that seems to be quite distinct from the normal spiral structure so clearly seen in optical emission.

Figure 7.2 The spiral galaxy NGC 4258 with the normal spiral arms outlined by bright stars and the anomalous arms observed in the radio continuum and X-rays. (NASA/CXC/University of Maryland/A.S. Wilson et al.; Optical: Pal.Obs. DSS; IR: NASA/JPL-Caltech; VLA: NRAO/AUI/NSF.)

Van der Kruit and collaborators interpreted these anomalous arms as being due to a recent ejection from the nucleus of the galaxy – an explosive ejection in two opposite directions more or less in the plane of the galaxy. These clouds encounter the ambient rotating interstellar medium which bends the ejected material to the form of a spiral; shock waves form along this spiral and compress the magnetic field that is always present in the disks of spiral galaxies. This enhances the synchrotron emission and the phenomenon of the anomalous radio arms is explained. To achieve this, 100 million solar masses of gas must be ejected with velocities up to 1600 km/s. The total kinetic energy of this hypothetical event would be in excess of 10^{57} ergs or 1000 solar rest masses – quite an energetic event indeed, but still 1000 times less energy than that required in powerful radio galaxies.

The Leiden group went further and suggested that such a mechanism may be generally responsible for the spiral arms in disk galaxies. There is a historical problem with spiral structure. As had been shown by Oort (Figure 7.3) decades earlier, the Milky Way, and indeed all disk galaxies, rotate differentially; that is to say, the rotation about the center is not solid body rotation like a phonograph record but the period of rotation is shorter at smaller radii than at larger radii. This leads to the wrap-up problem of spiral structure: if the arms are actually material structures, they should wrap up into very tightly wound structures in a few hundred million years – much shorter than the 10 billion year lifetime of galaxies. Then why do we observe spiral structure at all, in particular, very open spiral structure with very few windings of the arms?

There were two suggested solutions to this problem: first, the arms are not actual material arms but waves in the disk of gas and stars; the waves rotate at

Figure 7.3 Jan Oort at his (very extended) prime. Not only was he the unquestioned overlord of Dutch astronomy with many administrative duties, but he also remained intensely involved in science, especially phenomena occurring in the nuclei of galaxies. (Copyright, Leiden Observatory.)

their own rate and stars and gas pass through increasing their density as they do. Leading proponents of this idea were C.C. Lin and Frank Shu of the Massachusetts Institute of Technology.

But a second more radical idea was considered by the Leiden group on the basis of the observed anomalous arms in NGC 4258: large masses of gas, approaching 100 million solar masses, may be ejected in opposite directions from the nuclei of galaxies every billion years or so. This gas would be wound by differential rotation into a spiral form, but before disappearing completely another ejection would occur and regenerate the spiral structure. The essential problem with this idea, which Oort's group emphasized, is the source of such a large mass of gas; they estimated that this is 1000 times more gas than could be deposited within a billion years within the inner central region of NGC 4258 by normal processes of stellar evolution – winds, planetary nebulae, supernovae.

It would appear that to explain this mass requirement, one would be driven to more radical ideas, such as those suggested by Ambartsumian: that active galactic nuclei (long before the term came into general use) play a dominate role in determining the structure and evolution of galaxies through the ejection of large masses of gas and relativistic particles. The origin of this gas at the center was unknown, and Ambartsumian speculated that there must be a large "nonstellar mass" at the centers of galaxies. In his reliance on new and unknown physics, Ambartsumian's point of view was extreme, but, as we shall see, in a modern context he was not entirely wrong, at least not about the presence of a nonstellar mass in nuclei and influence of these nuclei on the evolution of galaxies.

What better environment to test these ideas than our local neighborhood, the center of the Milky Way, where large-scale gas features had actually been observed moving away from the center? The original expanding arm of Rougoor and Oort, the 3-kiloparsec arm, can be quite well modeled in terms of a ring of gas which is rotating and moving away from the center. That is to say, the observed dependence of the velocity of this feature (measured in the 21-cm line of neutral hydrogen) with angular distance from the center is entirely consistent with a symmetric ring of gas expanding away from the center with a velocity of 53 km/s and rotating with a velocity of 200 km/s. This suggests that perhaps a more symmetric explosion at the center, like a supernova but much more powerful, might be able to excite gas motions at this large distance from the center; it may not be necessary to shoot clouds out of the center in well-defined directions, as cannonballs with a well-aimed cannon. The possibility of a symmetric explosion was an idea that Kevin Prendergast and I considered as part of a general program on the hydrodynamic effects of explosions in gaseous disks.

We found that it was possible to produce an expanding ring of gas by such an explosion in the center of a differentially rotating gas disk (Figure 7.4). The ring

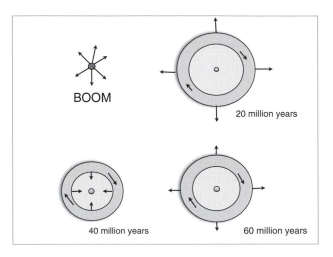

Figure 7.4 A massive explosion in the center of a symmetric galaxy can produce a ring of gas that oscillates radially inward and outward with a period of about 40 million years. This could cause features such as the observed 3-kiloparsec arm if observed in an outward moving phase.

first expands, sweeping up the rotating ambient gas in the disk (20 million years). But then because of conservation of angular momentum the ring is not rotating rapidly enough to balance gravity, so it contracts overshooting an equilibrium position, rotating faster and faster (40 million years) until it strikes a centrifugal barrier and re-expands (60 million years). Thus the ring oscillates about an equilibrium position in radius – it moves in and out with a period of around 40 million years with the radial motion gradually damping away – so that the actual time a feature such as the 3-kiloparsec arm could be observed after the initial explosion could be several hundred million years; it is not necessary for us to exist within a cosmically short timescale of 20 million years after the explosion in order to have a reasonable chance of seeing such a feature.

The problem again is that the energy and mass requirements were huge. Because most of the energy is directed out of the plane of the disk, to produce such a pronounced effect so far from the center would require the isotropic ejection (no directional dependence) of more than 100 million solar masses with a kinetic energy of more than 10^{58} ergs (the rest mass of 10 000 suns). The energy requirements approach those of active galaxies, but more serious is the mass requirement. This goes back to the issue identified by van der Kruit and collaborators for NGC 4258: what is the source of this large mass in the center – a mass certainly 100 times larger than any object currently at the center? To have recurring explosions, necessary to give a reasonable chance of seeing such phenomena, a large mass must somehow flow into the center on an astronomically short timescale. This, in fact, makes the entire scenario seem quite implausible.

7.2   Gas Motion in Barred Spiral Galaxies

Gerard de Vaucouleurs (Figure 7.5) was a Frenchman transplanted to the wilds of Texas and an expert on the morphological classification of galaxies. It had been known for some time that there were two sorts of spiral galaxies: so-called "normal spirals" which, apart from the often seen bright spiral arms, appeared to be axially symmetric about the center of the galaxy (axial symmetry means that the structure is the same on any circle drawn around the center).

Barred spirals, on the other hand, have a lower degree of symmetry; they are bisymmetric with a cigar-like extension through the center (bisymmetry means that the structure is symmetric about two planes perpendicular to each other). de Vaucouleurs, in his reclassification of galaxies, first carried out by Edwin Hubble, placed more emphasis on the barred systems because he realized that a very large fraction of disk systems, perhaps 40%, contain at least small central bars (it is now thought that more than 50% of disk galaxies are barred). But de Vaucouleurs also realized that if the mass distribution in barred galaxies is nonaxisymmetric, then the gravitational field is also nonaxisymmetric and that stars and gas would not move on neat circular orbits as supposedly they do in non-barred galaxies.

Let's consider this in more detail. A disk galaxy like the Milky Way has two main components – stars and gas. In most galaxies, the stars are by far the dominant component; the total mass of stars exceeds that of the gas by more than a

Figure 7.5 Gerard de Vaucouleurs, who first suggested that the noncircular gas motions toward the Galactic Center may be due to noncircular motion in the gravitational field of bar. (AIP Emilio Segre Visual Archives.)

Figure 7.6 A non-barred spiral galaxy, NGC 3810. The mass distribution and the gravitational field is thought to be generally axisymmetric (rotationally symmetric about the center point) apart from the small perturbations caused by the conspicuous spiral arms. This is a Hubble Space Telescope image, NASA/ESA.

factor of 10. That means that the distribution of stars essentially determines the gravitational field of the galaxy. In a barred galaxy, the bar is a substantial structure of the stellar component of the galaxies and the resulting gravitational field is not axisymmetric; the gas component responds to this nonaxisymmetric field imposed by the bar, but it does not determine the gravitational field.

Stars and gas can move rather differently in a given gravitational field, and, in the Galaxy, it is gas that we observe by means of the 21-cm line of neutral hydrogen as well as various molecular lines. A general star orbit in a non-barred galaxy, with an *axisymmetric* gravitational field (as in Figure 7.6) can be as that shown in Figure 7.8. The stars can loop and make large excursions from a circular path. However, orbits of gas particles (clouds) cannot loop but must follow a more simple path – a circular path. This is because the gas is collisional; where the orbit loops the gas would self-collide, heat up, lose energy, and settle into a circular orbit; the gas dissipates its energy present in noncircular motion. So the gas must follow simple periodic nonlooping orbits, or, in a symmetric potential, a circular orbit. The point is that gas orbits are more restricted than star orbits.

But what about a nonaxisymmetric gravitational field as in a barred spiral galaxy (Figure 7.7)? Then there are no circular periodic orbits. It even becomes more complicated when we realize that the figure of the bar may rotate – the bar may tumble – like a cigar spinning end over end. In that case the mass distribution and gravitational field is time dependent, and one could reasonably ask if steady-state gas flow is possible. Will the gas flow change with time in a nonreversible way? This is actually a subtle question; it is possible for the gas to lose

Figure 7.7 A barred spiral galaxy, NGC 1300. The mass distribution and the gravitational field here are bisymmetric. (identical on a line drawn through the center). The deviations from axial symmetry in the region of the bar are very large and the gas flow cannot be on circular orbits. HST image, NASA/ESA.

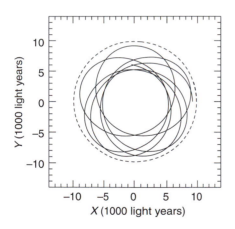

Figure 7.8 Possible star and gas orbits in an axisymmetric galaxy potential. The solid curve shows the possible path of a star in the Galaxy. The orbit can make large excursions from a circular path and it is not, in general, closed; it loops and does not return to its previous path. Gas, however, cannot loop because it is collisional; the gas will collide and lose energy. The only possible steady motion for the gas in an axisymmetric potential (unless it is disturbed by an explosion) is circular as is indicated by the dashed curve.

angular momentum to the more substantial bar in the stellar component and thereby flow inward on a timescale of several (on the order of ten) orbital periods. But, on a shorter timescale the gas motion can be reasonably steady when viewed in the *rotating* frame of the bar. Then to describe the gas motion we should look

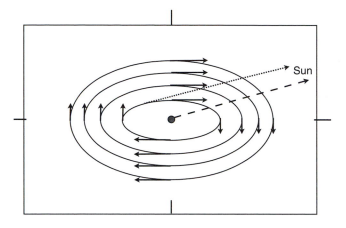

Figure 7.9 Periodic, orbits in a nonaxisymmetric (barred) gravitational field. The curves show a collection of orbits that gas could possibly follow in a barred potential. The massive bar is elongated horizontally and such orbits are possible only if the figure of the bar rotates; that is, the bar tumbles maintaining its shape. If the Sun is in the direction shown, an observer on Earth will observe apparent outflow in the direction of the center (dashed line). The observer will also see an apparent rotation velocity which is generally larger than that in a gravitational field created by the same mass with an axisymmetric distribution.

for simple periodic orbits in this rotating frame. It turns out that if the bar is tumbling rapidly enough there are simple nonintersecting periodic orbits that are elongated in the direction of the bar and are possible paths for gas, the so-called *streamlines*, as shown in Figure 7.9. For these elongated orbits to exist the tumbling period of the bar must be comparable to the periods of the outermost orbits.

The motion along such streamlines clearly deviates from circular, and, given the indicated sense of rotation, observers from the direction indicated by the dashed line would see apparent outflow, that is, motion away from the center of the galaxy. Moreover, the highest observed velocities from which a rotation curve is defined (along the dotted line) would give a false indication of the mass distribution if interpreted in terms of circular motion in an axisymmetric gravitational field; the "rotation" velocity would be too high. This can account for the strange peak in the rotation curve of Rougoor and Oort seen at a radius of 1800 light years (Figure 4.6); the peak is due not to an actual mass concentration but to the flow on elliptical streamlines. The significant point is that, such flow in a nonaxisymmetric gravitational field is a nonexplosive way of producing apparent expansion (or contraction) motion in the gas.

In 1963 the details of gas motion in barred spiral galaxies were not so well understood. But Gerard de Vaucouleurs did realize that in a nonaxisymmetric gravitational field the gas would move not on circular orbits but, broadly speaking, on elliptical orbits. Moreover, he had visual emission line observations of several barred spirals that indicated clear deviations from circular motion. So that year, at a meeting in Canberra, Australia, de Vaucouleurs proposed that the apparent expansion motion seen by Rougoor and Oort in the 21-cm hydrogen line observations in the direction of the Galactic Center could be just such noncircular motion in the gravitational field of a bar appropriately viewed – that there is no net outflow of gas from the center but a steady circulation of gas on such noncircular orbits – that there is no need to propose explosive events or ejections of large masses of gas from the nucleus to account for the observations.

At this point de Vaucouleurs' suggestion was the radical proposal, and, as is evident from the discussion following his talk, in particular the comments by Oort and Ambartsumian, it was not well received. In fact, for fifteen years his suggestion languished primarily because of a strong prejudice in favor of ejections or explosions occurring in galactic nuclei in general. But several developments changed this perception. First, there was a general realization that even in active galaxies, explosive or impulsive events play a less important role. Second, specific models of gas flow on elongated orbits were developed. In the years 1978 to 1980 the radio astronomers Butler Burton and Harvey Liszt published a series of papers interpreting the gas distribution and motions observed toward the center of the Galaxy as a bar-like structure tilted with respect to the Galactic plane as defined locally. The tilted structure accounted for van der Kruit's observation that the neutral hydrogen gas at forbidden velocities tended to lie above the plane on one side of the Galaxy but below on the opposite side. Moreover, in terms of this single unified model they were able to explain a number of the distinct features that, previously, required separate ejection events; thus the model had a distinct advantage of efficiency.

Their model for the gas distribution projected onto the plane of the sky is shown in Figure 7.10. This is actually an elliptical disk with an axis ratio of 3:1, oriented such that gas on such elliptical streamlines exhibits apparent outflow as seen by an observer at the position of the sun as shown in Figure 7.9. We notice that the elliptical disk (the bar) extends further on one side of the plane than on the other. This is simply because of our perspective in viewing such an elliptical disk from the position of the Sun; the bar is significantly closer to the Sun on the left side (positive angles, receding velocities) than on the right side, which gives the central structure a distinct asymmetric appearance. This, in a sense, is a prediction of the Burton–Liszt kinematic model. Ten years later, this effect was actually observed when near-infrared observations became so refined that the general distribution of stars in the central bulge of the Galaxy could be mapped.

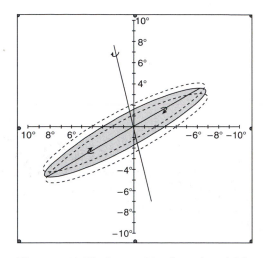

Figure 7.10 The Burton–Liszt barred model for neutral hydrogen motions observed toward the Galactic Center. This is the elliptical disk for the gas with an axis ratio of 3:1 and tilted by 26 degrees with respect to the plane of the Milky Way as defined locally. The long axis of the bar is oriented at an angle of about 20 degrees with respect to the line of sight to the center; the motion on elliptical streamlines gives apparent expansion motion away from the center and implies by perspective that the disk apparently extends further on the positive longitude side where velocities are receding (the left side as shown above) than on the negative longitude side. (Courtesy of Harvey Liszt.)

This is shown in Figure 7.11, which is a near-infrared map of the emission from old stars toward the central region of the Milky Way as observed by the Cosmic Background Explorer satellite (COBE) – the same satellite that discovered the fluctuations in the universal cosmic microwave background radiation. This remarkable picture shows clearly that the Milky Way is indeed a spiral galaxy seen edge-on from our vantage point in the plane, but also, on close inspection it demonstrates just such an asymmetry in the central bulge which should be identified with the bar. When we compare to the Kapteyn model of the Galaxy (the Universe!) eighty years earlier, this view of the Milky Way represents a truly remarkable human achievement!

As first emphasized by Leo Blitz and David Spergel on the basis of earlier infrared observations, this is striking confirmation of the existence of a bar in the Galaxy, a bar oriented just as required by the Burton–Liszt kinematic model for the observed gas motions.

In the years 1980–2000 there followed a number of papers supporting the bar model. It is important to realize that the Burton–Liszt barred model for the gas motion is not a consistent dynamical model. The gas streamlines are chosen to

Figure 7.11 The-near-infrared view of the Milky Way galaxy at large as produced by the COBE satellite. Careful inspection reveals just such an asymmetry predicted by the bar-like structure required to explain the noncircular gas motions in the central regions. (NASA, COsmic Background Explorer [COBE] Project.)

be ellipses of axial ratio 3:1 and oriented at a certain angle to the line of sight to best fit the observations; in this the model does very well. But these streamlines do not arise in a specific gravitational potential; the model is not dynamical. Although there had been earlier general studies of the motion of gas in barred potentials that demonstrated the relevance of periodic orbits (by myself and Jim Huntley, for example), the first detailed hydrodynamical calculations that actually traced the motion of gas in a realistic barred model for the Galaxy were carried out by Wim Mulder as part of his thesis work at Leiden University (1983). Mulder, in collaboration with B.T. Liem, produced a barred model of the Galaxy in which the 3-kiloparsec arm emerged as flow on elliptical streamlines as shown in Figure 7.8. Many other features of the neutral hydrogen distribution and motion were also reproduced in this model which required a material bar with a moderate axial ratio of 4:3 (major to minor axis), with the bar major axis being about 20 degrees from the line of sight to the Galactic Center. (Note that is the elongation of the mass distribution that forms the gravitational potential; the gas orbits that respond to this potential are more elongated, 3:1 as in the Burton-Liszt kinematic model.)

The gas density distribution resulting from flow in such a potential is shown in Figure 7.12. It is very interesting that the gas responds to such a weak bar in the gravitational field by forming a two-arm spiral. This suggests that bars may play an important role in exciting and maintaining spiral density waves.

A second group led by James Binney at Oxford (1991) aimed at describing the Scoville and Kaifu et al. expanding molecular ring (at a radius of 800 light years) in terms of periodic orbits in a rotating bar potential. They showed that, in general, the pattern of velocities observed in the molecular gas could be well traced by assuming that the gas streamlines are described by such periodic orbits. However, a problem with explaining both the molecular ring and the neutral hydrogen

Figure 7.12 Hydrodynamical calculations for the distribution and motion of gas in a barred model of the Galaxy by Mulder and Liem (1986). The dark regions indicate higher density. It is important to realize that, although the mass model for the Galaxy is barred with the long axis oriented horizontally, the gas response is spiral due to crowding of elliptical streamlines onto periodic orbits of different orientation. Noncircular motions exceeding 50 km/s, as in the 3-kiloparsec arm, are easily reproduced in this model. (Courtesy of Wim Mulder.)

gas motions further out (the 3-kiloparsec arm, for example) with the one bar is the rate at which the figure must rotate. In order to produce the elliptical orbits that are sufficiently elongated along the major axis the bar, must tumble with a period characteristic of the orbit times in the region where the gas streaming is observed. This means that the bar producing the noncircular gas motions of the molecular gas near the center must tumble almost three times faster than the bar, producing the motions accounting for the 3-kiloparsec arm further out. Perhaps there are two bars tumbling at different rates, but then the entire picture becomes more complicated and less satisfying. These are generally considered to be details to be worked out by more careful modeling.

Certainly by the turn of the millennium the idea that the Milky Way is a barred galaxy and that this accounts for most of the noncircular gas motions observed in the central region had become the paradigm. The older models based on ejections and explosions now seem terribly contrived and artificial. Why should we evoke explosive phenomena of unknown cause involving very large masses and energies to explain an observation – noncircular gas motions – which follows naturally from steady flow in a nonaxisymmetric gravitational field? The explanation based upon a barred mass distribution and resulting gas flow is particularly compelling when one realizes, as did de Vaucouleurs fifty years ago, that this sort of structure is present in more than half of disk galaxies.

7.3   Controversy in Astronomy

In 1962 Rougoor and Oort discovered significant noncircular motions in the neutral hydrogen gas in the inner Galaxy. These motions were generally in the sense of expansion – moving away from the Galactic Center. One decade later, Scoville and Kaifu et al. found higher apparent expansion motion in the molecular gas much closer to the center. In both cases the motion could be described by rotating rings or partial rings of gas expanding away from the Galactic Center.

It is now difficult to imagine that an explanation as contrived as explosions or ejections could have been invoked to explain these observations. But in fact, at the time this appeared to be completely consistent with eruptive phenomena thought to be seen in other galaxies – Seyfert galaxies, radio galaxies, M82. Because of this perceived consistency, the idea seemed to be most natural explanation.

Naturalness is an enormously important criterion when judging competing explanations for astronomical phenomena but one that is difficult to define or quantify. Certainly naturalness involves efficiency of explanation (Occam's razor): How many separate interventions are required to explain a particular set of observations? Does each kinematic feature in the gas distribution toward the Galactic Center require a separate ejection? Or can flow in a single barred model account for a number of apparently distinct features? From the work of Burton and Liszt, it would certainly seem so.

In astronomy, naturalness also involves reliance on known physics: is it necessary to fall back on completely new or poorly understood physical processes to explain the phenomenon? One of the strongest arguments against the plausibility of the explosive model is in fact provided by calculations, such as those cartooned in Figure 7.3, that demonstrate the extreme energy and mass ejection requirements for events that could excite motions at such large distances from the Galactic Center. One is pushed toward bizarre physics to explain such phenomena. From the present point of view, far more natural is explanation in terms of steady gas flow in a nonaxisymmetric gravitational field, generated by a bar in the mass distribution, the sort of mass distribution that is observed to exist in at least half of disk galaxies.

Was then eruptive explanation a dead end street? Did it lead astronomers astray for twenty years? I would argue that it did not – that, in fact, it motivated interest in the general phenomena of gas motion in more general gravitational fields, not just axisymmetric, and led directly to the hypothesis of the Galaxy being a barred system. This then is the value of controversy: it provokes – motivates – researchers to look for alternatives that may be more sensible or natural.

Were people like Victor Ambartsumian completely wrong and unhelpful when they proposed that activity in nuclei has a major effect on the evolution and structure of galaxies? Ambartsumian's viewpoint was certainly extreme, but he was correct in unintended respects – or at least his ideas are consistent with several features of the current paradigm on the evolution of galaxies. Most galaxies do seem to contain a nonstellar central mass; in extreme cases, this object can dramatically influence the evolution and structure of the galaxy by blowing gas away and stopping star formation – a process now known as feedback which is thought to limit the number of massive galaxies.

So old ideas return in a different guise and with a different vocabulary. But overall, new directions and new paradigms are born of controversy – the wellspring of scientific progress.

8

The Evolving View of Active Galactic Nuclei

8.1 The Jet Set

In the late 1970s, confronted by a more plausible explanation for noncircular gas motions observed toward the center of the Galaxy, the idea of impulsive ejections or explosions was falling out of favor. At the same time, the nearest example of an "exploding galaxy," M82, appeared to be a less dramatic phenomenon – spectacular but not a single energetic event. Reinterpretation of the geometry of the filaments above and below the plane of M82 suggested two expanding bubbles and not motion confined to be parallel to the rotation axis of the galaxy – almost perpendicular to the line of sight. This means the true three-dimensional velocity of the filaments is closer to the observed line-of-sight velocities of 200 to 300 km/s rather than several thousand kilometers per second as originally thought; then the kinetic energy in gas motion is almost 100 times lower than in the original model. In addition, accumulating observational evidence of furious star formation in the plane of the M82 (such as the presence of a number of remnants of massive short-lived stars – supernova remnants) suggested that the outward moving filaments were in fact a large-scale galactic wind fueled by a burst of star formation in the disk of the galaxy – certainly not a massive single explosion.

Even for obviously active galaxies, radio galaxies, the original model of sudden expulsions of enormous clouds of ionized gas and relativistic particles was severely challenged by a new model – a model involving the more steady ejection of relativistic particles from the nucleus to the radio emitting lobes – the jet model.

The model has a basis in historical observations. In 1917, Herber Curtis (who later participated in the great debate against Shapley) was observing the elliptical

Figure 8.1 The nucleus of the nearby elliptical galaxy M 87 and its visual jet. The jet extends 5000 light years from the center and is polarized. (NASA and the Hubble Heritage team [STScI/AURA].)

"nebula," M87. He discovered that, as in other such nebulae, the light distribution was smooth (no spiral structure); however, unlike in most other elliptical nebulae, there appeared to be a faint linear structure extending over a distance of 30 to 40 arc seconds: "a curious straight – ray apparently connected with the nucleus by a thin line of matter" wrote Curtis (see Figure 8.1).

Thirty years later, M87 was identified with the strongest radio source in Virgo (Virgo A); it is a radio galaxy and one of the very nearest. It was also soon discovered that the optical emission from Curtis' jet (by then known to extend at least 5000 light years) was polarized, and this fact, in 1964, led the Soviet astrophysicist Iosif Shklovsky to propose that the emission process is synchrotron radiation – relativistic electrons spiraling in a magnetic field. Shklovsky also pointed out that that if a jet consisted of relativistic electrons moving near the speed of light primarily in the direction of the jet, then they would beam their radiation in the forward direction, as the headlights of an automobile (an effect known as Doppler boosting). This means that if there are actually two jets moving in opposite directions from the nucleus but near the line of sight, then we would observe only the approaching jet; the receding jet would point its headlight away from us.

The discovery of quasars the year before revealed that the nearest such object, 3C 273, also exhibited a linear jet, extending over 200 000 light years away from the bright star-like object (see Figure 5.2). This was an early indication that we are observing the same phenomenon as in M87 but on a vastly larger and more energetic scale.

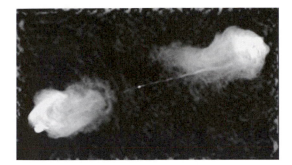

Figure 8.2 A high-resolution map of the radio galaxy Cygnus A at 6 centimeters. Here we get a different impression than given by the earlier radio maps such as that by Ryle et al. (see Figure 5.4). We now see that two jets appear to connect the nucleus of the visible galaxy with hot spots in the two radio emitting lobes. (Courtesy of VLA/NRAO, R. Perley et al. 1984.)

But what do jets have to do with the large extended radio structure in the powerful radio galaxies? The early radio maps (see Figure 5.4) of radio emission from sources such as Cygnus A gave the impression of two enormous lobes of relativistic particles and magnetic field expelled in opposite directions, perhaps by an energetic explosion. But in the late 1970s, when radio interferometers such as the Very Large Array (VLA) came on line, the much higher resolution radio maps, as that in the updated radiogram of Cygnus A shown in Figure 8.2, favored a quite different interpretation.

In the continuum radio emission we see two fine linear structures that connect the nucleus with "hot spots" in the large radio-emitting lobes. This certainly gives the impression of jets supplying the radio source with energy more or less continuously rather than a single energetic ejection. Here we can identify three essential aspects to a radio galaxy: the power source, which is certainly a compact region in the center; the pipe lines (the jets), which transport the energy to the exterior; and the hot spots, where the energy from the jets is deposited before diffusing out into the extended lobes. This is the essence of the jet model proposed *before* structures such as that evidenced by Cygnus A were actually observed; it is a theoretical prediction verified by later observations.

8.2   The View from Cambridge

The jet model was born out of perceived physical problems with the older model of lobes of gas and relativistic particles explosively ejected from the nucleus. A basic problem is that of adiabatic loss: if one allows a gas to expand into a large volume, it cools – this is called adiabatic cooling. In the explosive model of radio galaxies this is quite devastating because the clouds expand from a tiny nuclear region, perhaps less than a few light years, to a vast region of more than 100 000 light years. One might expect the adiabatic losses suffered by the emitting relativistic electrons to be substantial over such a large expansion factor. How do the electrons remain energetic enough to emit the observed radio emission by the synchrotron process?

But there was a second equally serious problem that became apparent when the hot spots were discovered at the Cambridge 5-kilometer interferometer (another project of Martin Ryle) in the early 1970s. The hot spots in Cygnus A are very intense regions of synchrotron emission; in fact, so intense that the relativistic electrons radiate away their energy very quickly; their lifetimes (the period of time over which they remain relativistic) are very short (less than 60 000 years) – much shorter than the light travel time from the nucleus of the parent galaxy (several hundred thousand years). This would seem to require continual re-supply of the relativistic particles from the center, most probably by a relativistic beam or jet.

And so, in 1974, the jet model was presented, primarily by Peter Scheuer and by Roger Blandford and Martin Rees at Cambridge University. Scheuer emphasized the problems presented by adiabatic loss and the appearance of the hot spots, and how a continuous beam of particles from the galaxy impinging upon the ambient medium can solve those problems. Blandford and Rees described a complete jet model, from launch to lobes. The model consists first of all of an energy source at the center, the source of outflowing energetic particles. Blandford and Rees did not commit themselves on the nature of the energy source but suggested that it could be a cluster of pulsars or a supermassive object or black hole. The jet is formed and focused hydrodynamically in a flattened gas distribution about the power supply, and, after exiting the nuclear region in two opposite directions, it drills its way through the external intergalactic medium. The end of the drill, where the jet meets the undisturbed medium, is called the working surface; that is where the particle velocity, previously directed along the jet axis, becomes randomized in a shock wave (particles then move in all directions), and the relativistic electrons efficiently emit synchrotron radiation as evidenced by the hot spots. The particles then diffuse out into the medium at large and form the large scale radio-emitting lobes.

We know now that jets are formed in the presence of disks about central objects in various astrophysical environments and on a wide range of scales – from forming stars and remnants of dead stars in the Galaxy to the powerful extragalactic radio sources. Matter accreting onto a central object will, because of its angular momentum, fall into a flattened disk, and there appears to be something very natural about the formation of jets by such disks. There is accretion through the disk as well as outflow, driven by various mechanisms (such as radiation pressure from a hot disk), along the rotation axis of the disk. But angular momentum is conserved, so the accreting matter transports not only mass but also angular momentum to the central object; in the case of a galactic nucleus that is presumably a black hole.

It had been appreciated for some time that black holes can possess the attribute of angular momentum – that they have spin and a rotation axis. In 1963, Roy Kerr, at the University of Texas, had found an exact solution to Einstein's field equations for the case of an axisymmetric, rotating black hole (because of the additional spatial variable, the Kerr solution is much more complicated than the Schwarzschild solution). For black holes formed entirely by disk accretion a very substantial fraction of the rest mass energy can be in the form of spin. The angular momentum of the black hole is a vector and establishes a definite direction in space; that direction can be only slowly changed – on a timescale comparable to that required to double the mass of the black hole by accretion. Matter accreting with an angular momentum vector pointing in a different direction than that of the black hole will, because of a relativistic effect known as "frame dragging," be forced to align its angular momentum to that of the black hole; the accretion disk will be forced into the plane perpendicular to the rotation axis of the black hole.

So the rotation axis of a black hole probably establishes axis of the jet. This is evident from radio sources such as NGC 6152 shown in Figure 8.3 where the jet axis has been stable for at least three million years – the timescale for the jet, if relativistic, to traverse the scale from the nucleus (a few light years) to the maximum extent of the radio source (three million light years). If the black hole angular momentum establishes the jet direction then it is also clear that in such objects the jet is formed very near the event horizon of the black hole. Very possibly the vast rotational energy of the massive black hole is being extracted to drive the jet by a process known as the Blandford–Znajek mechanism involving a tenuous magnetized accretion disk; in this case the disk does not shine brightly but allows the black hole to directly deposit its rotational energy electromagnetically into a two-sided jet. In other cases the disk, heated by the friction of accretion and forced by frame dragging to align itself perpendicular to the black hole rotation axis, provides a funnel for focusing the hot matter driven off the disk into a narrow jet. In either case the efficiency of converting the accreted matter into energy, luminous or kinetic, can be as high as 20%.

Jets – especially relativistic jets with the particles moving along the jet near the speed of light – appear to go together with accretion onto massive black holes. This is a natural explanation (i.e., relativistic beaming) for the often seen one-sidedness of jets in the enormous radio galaxies, such as NGC 6152. But highly relativistic jets can also account for a phenomenon first observed in 1970 after

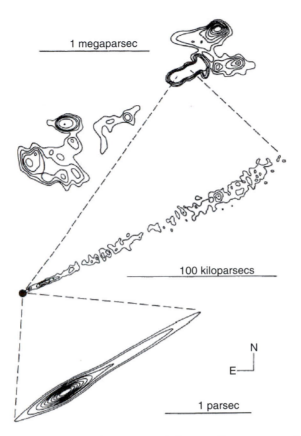

Figure 8.3 This is an extremely large radio source associated with the galaxy NGC 6251; the structure stretches over three million light years. The radio jet that emanates from the nucleus of the galaxy is shown on various scales, observed by different techniques. The remarkable aspect is that the direction of the jet is very precisely maintained over a range of one million in scale. Note that the jet appears to be one-sided, suggesting relativistic motion and significant beaming along the line of sight. If so, then the true size is actually much larger than the projected size of millions of light years. (Presented by Readhead et al. 1978.)

"very long baseline interferometry," VLBI, had been developed. VLBI is a technique of combining radio telescopes with continental or even intercontinental baselines as one huge radio telescope. With this technique, sources as small as a few light years can be resolved in distant quasars. When VLBI was first applied to observe quasars that appeared as compact or unresolved radio sources to single dish radio telescopes, it was discovered that the sources comprised knots or bright spots in light year scale radio jets. But the real surprise was that, when observed over intervals of several months, the bright knots appeared to move along the

Figure 8.4 Martin Rees, one of the most influential astrophysicists of the past half century. Rees has been a prolific producer of ideas that have turned out to be correct, among them the early prediction of apparent superluminal (faster than light) motion of radio bright spots moving relativistically near the observer's line of sight.

jet faster than the speed of light! (This observation gave a temporary boost to those supporters of non-cosmological redshifts who argued this was proof that the quasars were really closer than the redshift indicated.) But the effect had actually been predicted several years earlier by Martin Rees, who wrote in the journal *Nature* that "an object moving relativistically in suitable directions may appear to a distant observer to have a transverse velocity much greater than the velocity of light" (1966, Nature, Vol. 211, p. 469).

Not all jets associated with nuclear black holes are relativistic, nor does a black hole always produce an obvious jet; there are powerful active galactic nuclei (the radio quiet quasi-stellar objects, for example) with no conspicuous jets. It is also true that the presence of a jet does not indicate the presence of black hole; the jets formed around newly forming stars, for example, imply that the critical aspect is an accretion disk that focusses the outflow into a narrow jet. But a jet emerging from an active galactic nucleus would appear to be a strong indication of the presence of a massive black hole.

This is evident in many Seyfert galaxies. Unlike the large and powerful radio sources, usually identified with elliptical galaxies, the Seyfert phenomenon is normally associated with spiral galaxy hosts. One such nearby Seyfert galaxy, NGC 1068, is shown in Figure 8.5. Apart from its bright stellar-like nucleus it appears to be quite a normal spiral galaxy. In the early 1980s, when the radio astronomers began observing Seyfert galaxies with high resolution using the VLA, this active galaxy was found to exhibit what, morphologically, appeared to be a radio galaxy,

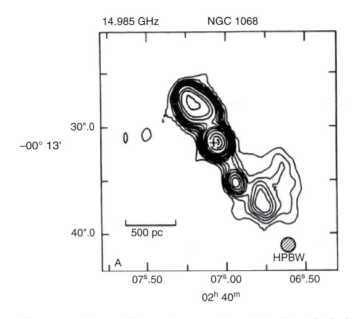

Figure 8.5 A map of the continuum radio emission (2 cm) in the inner region of Seyfert galaxy NGC 1068. The indicated scale is 500 parsec, which corresponds to about 1500 light years, so the structure is within the visible image of the galaxy. The cross is the optical nucleus of the galaxy. This is reminiscent of the much larger and more powerful structures in radio galaxies. Although the size of the jet, at about 4000 light years, is vastly smaller, the jet structure is essentially the same. The observations were made by Andrew Wilson and James Ulvestad in 1982 using the VLA (NRAO).

but on a much smaller scale. This structure is shown in Figure 8.5 where we see a linear structure stretching over 4000 light years in the central region. On one side, there is even a hot spot indicating the interaction of the jet with the ambient medium, in this case, the interstellar medium within the galaxy. Many Seyfert galaxies reveal such radio morphology in their inner regions, which suggests that the basic mechanism, the power source, is disk accretion onto a massive black hole. The bright optical radiation seen in Seyfert galaxies, the star-like nucleus, almost certainly arises from the accretion disk, heated by the friction between turbulent elements as described in Chapter 5.

But what about normal galaxies such as the Milky Way or nearby galaxies not classified as "active"?

8.3   A Massive Black Hole in NGC 4258

The mechanism for producing powerful jets and ensuring their stability is most likely an accretion disk about a massive black hole. So the presence of a jet is suggestive of a black hole but by no means proves that a black hole is present; that proof requires dynamical evidence, which is to say, observations of the motion of stars or gas within a region of well-determined size. The first such definite evidence was provided by the galaxy, NGC 4258 – that very same system with the anomalous radio-emitting "spiral arms" discovered by van der Kruit, Oort, and Mathewson at Westerbork in 1972.

In the early 1980s there was a general reconsideration of the anomalous arms in terms of the jet model. The arms were seen to be consistent with a jet flowing from the nucleus of this galaxy, but near the plane of the disk and bent either by the rotating interstellar gas or by a precession of the jet itself. This offered a viable alternative to the impulsive ejection of massive clouds; again the process was continuous and more gradual, involving a moderate mass loss rate of 0.001 to 0.01 of a solar mass per year. Thus in an otherwise normal galaxy there is a jet that suggests that the same sort of power supply evidenced by active galaxies is present in normal galaxies: a massive black hole. But definitive proof of a black hole came in 1995 through a truly remarkable observation.

Earlier, in the mid-1960s, the rapid development of radio receivers at various wavelengths for radio astronomy led to the discovery that spectral lines of certain molecules, such as hydroxyl radical (OH, an oxygen and hydrogen atom combined in a electrically neutral form) and water vapor (H_2O), were amplified in special astrophysical environments – they were present as masers ("microwave amplification of stimulated emission").

This requires a bit of explanation. Spectral lines arise from transitions between energy levels in atoms or molecules. There is an upper energy level and a lower energy level associated with a particular spectral line. When an isolated molecule makes a transition from the upper to the lower level, a photon with an energy corresponding to the energy difference in the two levels is emitted. This is called spontaneous emission because it can happen spontaneously with some probability. A molecule in the lower energy level can also absorb a passing photon with the right energy to cause a transition from the lower to the upper level. The photon, in effect, disappears, and that is called absorption. But there is another possibility: the passing photon with an energy equal to that of the transition can also stimulate a transition from the upper level to the lower level, and that is called stimulated emission. Then the passing photon is joined by a second photon; one photon becomes two photons of the same energy or wavelength.

Under most conditions stimulated emission is less important than absorption, and radiation from a background source at the energy of the transition is scattered or dimmed by a medium containing that molecular species; the distant observer sees an absorption line. But that is not always true in the interstellar medium where the gas is very tenuous and the populations of the two energy states in the molecules do not have their normal ratios – when the upper level of the transition is *overpopulated* relative to the lower level. Then stimulated emission can win.

Two conditions are necessary for this to happen: first, there has to be a nearby source of radiation with a different wavelength, visible light for example, to cause the relative overpopulation of the upper state – to pump up the higher energy state. This can happen, for example, when the pumping radiation preferentially causes a transition from the lower state to a third much higher energy level and, in effect, depopulates the lower level. Second, the change in gas velocity over the region where the line is created must be very small, so that a photon emitted at the wavelength of the transition can interact with the same transition in other molecules along the line of sight – that the transition is not Doppler shifted away by relative motion of molecules. Then one photon becomes two, which become four, which become eight, and so on; it is this cascade of photons, all with the same wavelength and all moving in the same direction, that produces the enormous amplification of the spectral emission line.

Maser emission from water vapor occurs in a spectral line with a wavelength of 1.35 centimeters, and it appears as intense emission. In the Galaxy, this occurs in small regions near newly forming stars; it is the young luminous stars that provide the infrared or optical radiation that pumps the upper level of the water vapor transition relative to the lower level. The total power in the single narrow spectral line can be as high as one-tenth of the Sun's luminosity. Masers are also seen in star forming regions in other galaxies, but in the early 1980s very powerful masers, called megamasers, were discovered in the nuclear regions of several nearby galaxies including NGC 4258. For these sources, the luminosity in the spectral line can be as high as several hundred times that of the Sun.

The combination of high brightness and small size means that these masers are ideal sources for observing with VLBI. Moreover, the fact that the emission is in a spectral line means that the relative velocities of emission regions along the line of sight can be determined by means of the Doppler shift. In the mid-1990s this technique was applied to the water maser sources in the central region of NGC 4258 and yielded the first convincing evidence for a massive black hole in a galactic nucleus.

The results were published in the journal *Nature* in 1995 by Makoto Miyoshi, James Moran, James Herrnstein, Lincoln Greenhill, Naomasa Nakai, Phillip Diamond, and Makoto Inoue (observations using VLBI techniques frequently have a large number of co-authors because of the number of separate observing stations involved; this paper is not exceptional in this respect). They found that the maser sources appeared to be in a disk or ring with an inner radius of 0.4 light years and an outer radius of 0.8 light years. The motion was consistent with rotation about the very center of the galaxy.

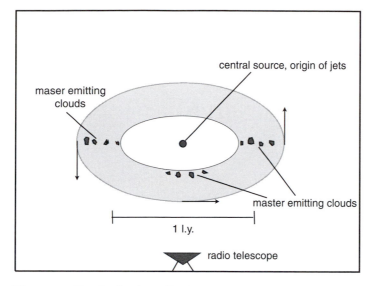

Figure 8.6 The distribution of individual water masers in the central region of NGC 4258 (schematic). They appear to be in a slightly warped annulus rotating about the center of the galaxy but oriented almost perpendicular to the plane of the galaxy itself. The jets connected with the anomalous arms emerge from the center of this nuclear disk and are oriented perpendicular to it. This places the jet nearly in the plane of the larger galaxy. The water vapor maser sources are located such that amplification takes place for an observer in the indicated position. The masers on the receding and approaching side of the disk exhibit Keplerian motion about a point mass and those in the direction of the center show the linear velocity gradient expected for a disk with a central hole (see Figure 8.7).

The distribution of these sources in the disk is illustrated in Figure 8.6, where we see that they appear to be near major and minor axes of a very inclined disk. In fact, this is where the variation in velocity along an inclined disk seen in projection is lowest – the small change in velocity necessary for a maser. It is this nearly edge-on orientation of the disk that allows the observer to see the maser amplification of the water vapor line; the amplification occurs in the disk itself. The pumping radiation is most certainly from the central object.

The velocity field of these maser regions is most revealing. The rotation as a function of angular distance to the disk is shown in Figure 8.7. The maximum velocity at the inner edge of the disk is in excess of 1000 km/s. The maser regions very nearly in the direction of the center, toward the continuum source, have a very low velocity with respect to the galaxy because their motion, being rotation about the center, is essentially perpendicular to the line of sight; the linear dependence of line-of-sight velocity with distance from the center is consistent with circular motion projected onto the plane of the sky.

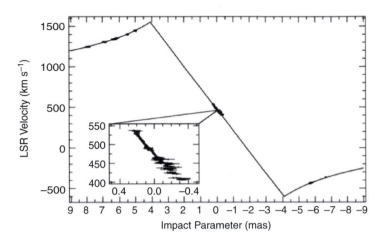

Figure 8.7 The velocity of water masers plotted against angular distance in milli-arc seconds (1/1000 of an arc second) from the center of NGC 4258. The curve is the predicted velocity field for a point mass of 37 million solar masses. From Miyoshi et al. 1995, with permission of M. Miyoshi.

Beyond the inner edge of the disk the speed falls as the inverse square root of the distance from the center – exactly as the orbital speed of planets does in the Solar System, exactly as Newtonian gravity predicts for motion about a point mass. There is a point mass, or at least a spherical mass with a radius less than 0.4 light years within this rotating ring. And the value for that mass turns out to be 37 million times larger than the mass of the Sun. This would correspond to an average density of more than 100 million suns per cubic light year within the inner radius of the disk!

8.4   Are There Alternatives to a Black Hole?

Could this mass concentration, within one-half light year, be something besides a black hole? For example, could it be a dense cluster of ordinary stars lying entirely within the inner boundary of the disk? If there were such a cluster then the time for a single star to undergo a disrupting collision would be shorter than 100 million years. Therefore we would expect the individual stars in the cluster to disappear (and form a black hole) on very short order compared to the age of the galaxy.

Could it be a cluster of very small dense stars – stellar remnants such as neutron stars or stellar mass black holes? Then, because of the much smaller geometrical cross section of the stars, the collision time would be at least 10 million times longer than the age of the Universe and there is no problem with the neutron stars being destroyed through collisions. But there is another problem. Recall from Chapter 5 that a dense stellar system dynamically evolves through distant gravitational encounters between stars. Some stars are ejected, the remaining cluster contracts and becomes denser (see Figure 5.7) and in a finite time approaches the singularity of infinite density. For such a dense cluster of stellar remnants, this would happen in less than a billion years, again a much shorter time than the age of the galaxy.

Could the massive object be a more exotic configuration of subatomic particles resembling a neutron star, supported against gravity by degenerate or packing pressure? In fact, there are two general possibilities because there are two basic sorts of subatomic particles that are distinguished by their quantized spin (the unit of quantized angular momentum is $h/2\pi$, where h is the tiny Planck's constant). There are the fermions, such as electrons, protons, neutrons, neutrinos, with half integral spin. Such particles cannot occupy the same quantum state; they are subject to the packing or degenerate pressure mentioned in Chapter 1. And then there are bosons. Bosons are particles with integral spin such as photons or the carriers of the weak nuclear force, the Z and W bosons. An unlimited number of bosons can crowd into the same quantum state, the lowest energy state, so there is no packing pressure; there is, however, a minimum length scale associated with these particles – the Compton wavelength.

It is the packing pressure of fermions that supports white dwarfs (electron degenerate pressure) and neutron stars (neutron degenerate pressure) against gravitational collapse, but only if the object has a mass below a critical mass limit. It was demonstrated many years ago by Chandrasekhar and by Oppenheimer and Volkoff that this critical mass is less than two times the Sun's mass for such objects. It would certainly be impossible to construct a supermassive object supported by electron or neutron degenerate pressure. But there is another sort of fermion – the neutrino. Neutrinos are neutral (uncharged) subatomic particles that interact very weakly with other matter. It was once thought that they were completely massless, but it is now known that neutrinos do have a small mass, probably less than one electron volt (expressing mass in units of energy via $E = Mc^2$, an electron volt corresponds to a mass of 1.8×10^{-33} grams). Because neutrinos, like electrons and neutrons, are fermions, their velocity, and hence pressure, increases when too many are packed into a specific volume. Could we replace massive black holes by neutrino stars?

For a fermion object the critical mass, the mass above which the object would collapse into a black hole, depends on the inverse square of the mass of the fermion particle, in this case, the neutrino (i.e., $M_c \propto 1/m_v^2$). That critical mass should be at least ten billion times that of the Sun; that is the estimated maximum mass of the dark objects at the centers of very massive galaxies like M87 (see Chapter 11). We assume such a large critical mass because it would seem inelegant to propose that the object in NGC 4258 is a neutrino star but in more massive galaxies the central objects are black holes. If neutrino stars can exist with masses of ten billion solar masses, then the neutrino mass must be at least 10 000 times larger than the mass of the three ordinary neutrino types, 10 000 electron volts (10 kilo-electron volts or keV), so it would have to be some sort of nonstandard, undiscovered neutrino (such particles are within the realm of possibility).

Below this critical mass, there is a relationship between the radius and the mass that is valid for white dwarfs, neutron stars, or neutrino stars. The radius is proportional to the inverse cube root of the mass of the object, so the more massive the fermion star the smaller the radius. At the critical mass this radius becomes roughly the Schwarzschild radius of the object so the mass radius relationship is $R \approx R_{sch}(M_c/M)^{1/3}$, where R_{sch} is the Schwarzschild radius of the critical mass M_c. Assuming a critical mass of 10 billion solar masses, the radius turns out to be about 0.02 light years for an object of 37 million solar masses, as in NGC 4258. So the neutrino star could fit comfortably inside the inner radius of the gas disk in NGC 4258 (0.4 light years); to the disk the neutrino star would appear as a point mass, and it is possible that a neutrino star could be a black hole emulator in NGC 4258. (For more on supermassive neutrino stars see Viollier, Trautman, and Tupper.)

It is also possible to construct stable objects made of bosons – boson stars – but the known bosons (photons, Z and W bosons) are not candidates because they are unstable or charged or they have zero mass (e.g., photons). But there is the possibility of undiscovered bosons with zero spin, so-called scalar particles (the newly discovered Higgs boson is such a scalar particle but it is very short lived). As for all bosons, there are no constraints on their occupation of quantum states; there is no packing or degenerate pressure like that holding up white dwarfs or neutron stars against gravity. In this case one could say that the supporting pressure is a sort of quantum pressure arising from the Heisenberg Uncertainty Principle: it is not possible to pack bosons into a region smaller than that permitted by quantum physics. That critical length scale associated with bosons is called the Compton wavelength, and it is inversely proportional to the mass of the boson particle. The radius of the boson star cannot be smaller than the Compton wavelength, and, of course, if it is to be an alternative to a black hole then the Compton wavelength must be somewhat larger than the Schwarzschild radius. Equating the Compton wavelength to the Schwarzschild radius again defines a critical mass – an upper limit to the mass of the boson star if it is to be outside of its Schwarzschild radius; in this case the critical mass is inversely proportional to mass of the boson ($M_c \propto 1/m_b$). If the critical mass is 10 billion solar masses then the mass of the boson is an incredibly small 10^{-53} grams or only 10^{-20} electron volts.

For a boson stars there is also a mass radius relationship: the radius is proportional to the inverse of the mass ($R \approx R_{sch} M_c / M$, where R_{sch} is the Schwarzschild radius of the critical mass M_c). Again taking a critical mass of ten billion solar masses, we would find that for a 37-million solar mass object this turns out to be almost one light year, so the entire molecular disk in NGC 4258 would be deep inside the boson star (possible because the boson stars have no sharp edge). But the gravitational force in the disk, from 0.4 to 0.8 light years, falls off with radius, exactly as it should do for a point mass and this would not happen if it were embedded in the extended mass distribution of the boson star. This effectively rules out boson stars (at least of the simplest kind) as a substitute for the black hole. We will see later that the gravitational field about the massive object in the Galactic Center as probed by individual stellar orbits also rules out neutrino stars. (For more on boson stars see the review by Liebling and Palenzuela.)

In the context of standard physics, there seems to be little alternative to a massive black hole of 37 million solar masses in the center of NGC 4258; there remains only the remote possibility of a fermion ball in the form of a neutrino star. And NGC 4258 is a galaxy that, apart from the jets evidenced by the anomalous spiral arms, shows no spectacular activity, certainly not on the level of Seyfert galaxies. It is those jets that lead directly back into the massive black hole at the center. We see the jet, we see the disk, we infer the black hole.

8.5   Reflections

It is difficult now to appreciate the impact of the Blandford–Rees paper and the fact that it was successfully predictive: the paper appeared a few years before large-scale radio-emitting jets were actually observed. The model also put to rest the idea of large-scale explosive or impulsive ejections from the nuclei of galaxies; the process was smoother and more continuous.

That is not to say that controversy ended with the publication of the Blandford–Rees model in 1974. There was resistance by the proponents of explosive ejections certainly through 1976, and there were alternative models, such as that proposed by Bill Saslaw, Mauri Valtonen, and Sverre Aarseth. This model involved the two-sided ejection of massive objects, particle accelerators, into the intergalactic space; the hot spots were then interpreted as regions surrounding the *in situ* acceleration of relativistic particles by the massive objects. But the jet model was convincingly confirmed by the observation of large-scale radio jets in many classical radio sources. Rarely in astronomy has a model had such a spectacular success so soon after being proposed; it was an idea whose time had come.

The Blandford–Rees jet model was part of a general theoretical trend away from impulsive or explosive events in galactic nuclei. It was cogent because it was based on existent observations, such as the long-ago discovered optical jet in M 87 as well as the hot spots in Cygnus A, and shortly afterwards with observations that convincingly revealed that the radio-emitting lobes were supplied more or less continuously through jets by the power source in the nucleus.

Increasingly on theoretical grounds, that power source became identified with a massive black hole powered by accretion of matter through a disk, a natural configuration when one appreciates that any matter with even low angular momentum falling from scales of many light years down to a region of a few million kilometers will form a rapidly rotating disk. The disk is necessary for formation of the jet, either by providing a funnel for mass blown off the hot disk or by extracting the rotational energy of the black hole. Moreover, that disk will inevitably be forced into a plane perpendicular to the rotation axis of the black hole; so it is that axis that establishes the direction of the jet, and this can be quite stable over millions of years.

Then, remarkably, in 1995 that disk was observed in the nucleus of NGC 4258; its rotation velocity implied an object 37 million times larger than the mass of the Sun within a region of half a light year. If Einstein's General Theory of Relativity is the correct theory of gravity (at least in the limit of these very strong fields), there is almost nothing else (apart from a fermion ball) that this could be but a black hole surrounded by its accretion disk.

This all began to resemble Lynden-Bell's scenario for galactic nuclei as old quasars. But how general is the presence of black holes? NGC 4258 is quite a special case, with water vapor megamasers in a nuclear disk inclined almost perpendicular to our line of sigh. This special configuration must be rare in galactic nuclei. The observational technique, VLBI observations of compact maser regions, as powerful as it is, can be applied only in a very few cases. Are there black holes in the nuclei of other normal spiral galaxies? Is there a massive black hole in the center of the Milky Way as Lynden-Bell had suggested? The decline of the explosion model in favor of the barred spiral model for noncircular gas motions observed toward the Galactic Center and the absence of other signs of "activity" (no conspicuous jets, for example) weakened the case for a massive black hole in the Milky Way. But there was a further problem: the discovery of newly formed massive stars near Sgr A*; the "paradox of youth."

9

The "Paradox of Youth": Young Stars in the Galactic Center

9.1 The View from Groningen, 1983

In the late spring of 1983 the more observant residents of Groningen might have noticed groups of unusual dreamy-looking people wandering their streets and squares muttering among themselves about spirals, gas, stars, and bars (but not the usual sort which are in abundant supply in the town). They might have even heard the occasional mention of a black hole, although they probably would have thought that the visitors were discussing the large and unpleasant underground public toilet on the Grote Markt (the central market square).

Most Groningers do not realize (or much care) that their town is the birthplace of the modern study of the structure the Galaxy – that this is where, 100 years earlier, Jacobus Kapteyn began his measurements of the images of tens of thousands of stars on photographic plates and devised his model of the Milky Way – the Kapteyn Universe – which placed the Sun near the center of the great star system. It is the university town where Jan Oort defended his thesis in 1926 before moving on to the more cosmopolitan center of Leiden. And now Oort was back in town, along with about 200 other astronomers for an international symposium at the Kapteyn Institute appropriately entitled *The Milky Way Galaxy*.

This meeting attracted the most prominent astronomers who were working or had worked on various aspects of galactic structure and dynamics. Included among these were Donald Lynden-Bell, Martin Rees, Jerry Ostriker, and Martin Schmidt. There were even several eminent historians of science, Owen Gingerich, for example, who discussed various developments related to the discovery of the Milky Way as a spiral galaxy.

Jan Oort gave the review lecture on the Galactic nucleus, and it was a remarkably masterful and complete survey of where the subject stood at that time. With respect to the mini-spiral seen in the radio continuum in the central few light years (see Figure 6.5), Oort thought that this structure resembled jets; that the morphology was evidence for ongoing ejection from a central point source possibly from a rotating nozzle. He noted that the infall and tidal stretching of clouds is also a possibility but less likely, as it would seem to require three such clouds infalling at once and that the timescale for such infall is very short on Galactic scales – ten thousand to one hundred thousand years.

He remarked that the evidence for a massive black hole is not compelling – that the mass in the region could be accounted for by the cusp-like distribution of old stars and that the total luminosity from the inner 3 or 4 light years (not extraordinary at 10 million solar luminosities) as well as the state of ionization of the gas in the central few light years could be entirely explained by hot young stars formed in the core of the Galaxy about a million years ago. This idea had been suggested in 1982 by George Rieke and Marcia Lebofsky (Stewart Observatory at the University of Arizona) based on the presence of a number of giant red stars in the inner few light years. These are evolved massive stars (their structure has significantly changed over a short timescale) with lifetimes of less than a few million years. If these stars formed with a typical distribution by mass (the initial mass function), then there must be lower mass blue stars connected with this burst of star formation (although they had not at that point been identified), and these hot stars would be emitting the ultraviolet radiation sufficient to ionize the gas in the inner region. Oort's emphasis on a nuclear star burst was insightful given the discoveries soon to come.

With respect to ejections causing the expanding gas features at a large distance from the center, such as the famous 3-kiloparsec arm and the expanding arm at 135 km/s, Oort conceded that these are most likely due to noncircular motion on elliptical orbits produced by a bar, but he did not give up entirely on explosive or transient phenomena. He pointed out that most of the massive molecular clouds are on one side of the Galactic Center, which must be a highly temporary arrangement because the time for the clouds to spread out through Galactic rotation there should be very short, on the order of a million years, because of the short rotation period near the center. He interpreted the "expanding molecular ring" at several hundred light years (Scoville's and Kaifu's ring) as having an explosive origin, but (and this is a significant concession) possibly due to supernovae in the central region resulting from the recent burst of star formation, not to a single impulsive Ambartsumian kind of event.

He emphasized a then new discovery coming from high-energy gamma-ray observations – a spectral line at 511 kilo-electron volts (keV), the rest mass energy of the electron. This could result only from the annihilation of electrons and positrons, the antiparticle of electrons. Positrons have the same mass but a positive instead of a negative charge. When positrons and electrons encounter one another they form a short-lived bound atom called a positronium, but then, in a fraction of a second, this very unstable atom annihilates, producing two photons each having the rest energy of the electron 511 keV. Gamma-rays cannot penetrate the atmosphere of Earth but now this feature had been seen from the direction of the Galactic Center by a balloon-borne detector – at least sometimes seen. The line appeared to be transient; at times it was there, at time not. This apparent short time scale variability indicated a compact source of high-energy particles, and it had been suggested that the source was a small dense region with a high density of energetic photons where pairs of electrons and positrons could be produced before combining and producing the line. Oort took this to be the most convincing evidence for an unusual object at the center – a black hole. (It is now known that the annihilation line is extended over the region of the central bulge and the inner disk of the Galaxy and is due to the creation of positrons by cosmic rays as well as decays of radioactive isotopes produced in supernovae; it is not transient.)

Oort was at the time eighty-three years old (Figure 9.1), and this lecture marked, in a sense, his valediction to the study of the Galactic Center. He seemed to realize that the facilities did not exist in the Netherlands for precise infrared studies or high-energy observations and that this limited the contribution that could be made in a small country. Of course, the Netherlands was one of the founding members of the European Southern Observatory (ESO). This model for European scientific cooperation was becoming a major world center for research in astronomy with developing facilities, of the right kind, in Chile. But Oort, being retired for many years, no longer had the authority and power to push younger people

Figure 9.1 Jan Oort in his later years. Age did not dim the warmth and intelligent curiosity shining in his eyes.

into this subject. He spent the remainder of his long and active life contemplating the large-scale structure of galaxies in the Universe, writing, also in 1983, a major review paper on superclusters of galaxies. For seventy years his published work covered topics from Galactic dynamics to comets in the Solar System to neutral hydrogen observations of the Galaxy to the large-scale structure of the Universe. But his final publication at age ninety-two was a popular article for the journal *Mercury* on galactic nuclei, a subject that held him to the end of this remarkable life in astronomy.

9.2   A Nuclear Starburst

Compared to the compact cores of distant radio galaxies the radio source Sgr A* is not special; it would not be detected at the cosmic distances of radio galaxies or quasars; it is many orders of magnitude less powerful than these spectacular objects. But this tiny source near the center of extended distribution of old stars is certainly unique in the Galaxy. Its power in radio waves alone is larger than that emitted by twenty suns over all wavelengths, and it is 10 000 times more powerful than the other very compact radio sources in the Galaxy – the pulsars that are associated with spinning neutron stars. Its size is less than twenty astronomical units, the distance between the Sun and the Earth. The distance between the Sun and Earth. If there is an unusual object in the center of the Galaxy, a massive black hole, it would be natural to associate the object with this unique radio source.

Over the lifetime of the Galaxy, due to gravitational interactions between stars in the vicinity of the black hole, a large number of stars should collect in the region where the gravitational field of the black hole dominates; a peak in the stellar density, or a cusp in the radial distribution of stellar density, should form near the black hole. Thus we might expect to see a source of infrared radiation, the normal emission from old stars, associated with the cusp about Sgr A*. The brightest source of near-infrared emission at the center is infrared source (IRS) 7, but that is clearly emission from a luminous red star (a supergiant) lying a light year away from Sgr A*. The nearest source to Sgr A* is IRS 16, but this is a somewhat unusual source. Compared to the other infrared sources it is rather blue (that is, stronger at shorter wavelengths), and it is not particularly intense. As higher resolution observations became available, it was seen to break up into a number of individual components, at least nine. One might have hoped that at least one of these components would coincide with Sgr A*. Surprisingly, this turned out not to be the case.

Positional astronomy – the precise measurement of the positions of astronomical objects, absolutely and with respect to other objects – is the most basic and traditional form of astronomical observations. This is what Tycho Brahe did with respect to the motion of planets across the sky and it led directly to the Copernican revolution, Kepler's laws of planetary motion, and Newton's monumental description of the gravitational force. This is what Kapteyn did with respect to stars in the Galaxy and it led directly to Lindblad's and Oort's discovery of the rotation and the true scale of the Galaxy. So positional astronomy may seem tedious and dull, but it can give rise to the most significant discoveries.

In 1984 David Allen, then a young astronomer at the Anglo-Australian Observatory, observed the region near Sgr A* using the observatory's 3.9 meter telescope near Coonabarabran, New South Wales. His purpose was to locate the near-infrared sources with respect to the 6-centimeter radio continuum showing Sgr A* and the mini-spiral in order to find the infrared emission corresponding to the compact radio source, presumably one of the components of IRS 16. To do this he needed a positional accuracy of one-half an arc second or about 0.06 light years at the Galactic Center. Allen made maps at the near-infrared wavelengths of 1.6, 2.2, and 3.8-microns having a high positional accuracy, better than one-fifth of an arc second. Then he noticed that several of the compact features on the 3.8-micron map corresponded to features on the 6-centimeter radio continuum map. In this way he was able to precisely locate the 2.2-micron sources with respect to Sgr A* (see Figure 8.2), and found that none of the IRS 16 sources corresponded in position to the compact radio source. There appeared to be no near-infrared counterpart to Sgr A* – no evidence for a cusp of old stars about the putative black hole.

This had been suggested in earlier observations by Ron Ekers and his collaborators, but here was a certain proof. What then were the IRS 16 sources? Allen noted that they had infrared colors (after correcting for dust obscuration) of blue luminous stars, so he suggested that they could be associations of hot stars, and could provide the source of radiation that ionized the gas in the central few light years; there was no need for an additional source of radiation connected with a point source at the position of Sgr A* (see Figure 9.2).

This result had considerable impact on the proposal that there is a massive black hole at the Galactic Center. There was no evidence for a cusp of old stars and most of the radiation appeared to be provided by newly formed stars. But there was another quite devastating argument: if there were a black hole at the Galactic Center of one million solar masses or more, then it would exert strong tides on all of the material within the inner one or two light years, where the young stars are found. The effect of tides is to pull objects apart, just as the moon pulls out the oceans of the Earth on two opposite sides. The tide would act to prevent gas clouds from collapsing to form stars. In order for a gas cloud to collapse gravitationally it must have a large self-gravity to overcome the effect of

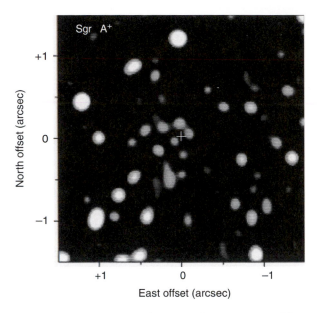

Figure 9.2 The position of nearby infrared sources with respect to the compact radio source Sgr A*, shown by the cross. The nearby objects are the young stars that comprise the components of the infrared source IRS 16. One second of arc corresponds to about one-tenth of a light year at the distance of the Galactic Center. This is from improved later observations by Menten et al. (1997).

tides. This places a constraint on the initial density of the clouds: to form stars within one-half light year of black hole of several million solar masses (one-half light year is the projected distance of the IRS 16 sources from Sgr A*), the initial density of the clouds must be in excess of 10^{11} particles per cubic centimeters – an exceedingly high-density interstellar material. There is no evidence for such dense clouds presently in the inner few light years.

So how can stars be formed so close to a massive black hole? An obvious solution is that there is no such massive black hole in the central region of the Galaxy. If there is no massive black hole, then there is no cusp and there are no tidal constraints on star formation. So the presence of young stars so very near the compact radio source, Sgr A*, appeared to call into question the very existence of a massive black hole. David Allen and I reached this conclusion in an interpretive paper where we placed an upper limit of several hundred solar masses on the mass of a possible black hole associated with Sgr A*. The state of affairs in 1988 was neatly summed up in a lecture given by Sterl Phinney (Caltech) at an international symposium on the Galactic Center at UCLA: given the low luminosity of Sgr A* over the entire electromagnetic spectrum and the nearby presence of young stars with sufficient luminosity to meet the energy requirements of the central region, the evidence for a massive black hole was hardly compelling. At the same time, however, dynamical evidence from the motions of gas and stars in support of a central nonstellar mass of a few million solar masses was accumulating.

9.3   Gas Distribution and Motions Near Sgr A*

In 1982, Eric Becklin, Ian Gately, and Michael Werner, using observations made more than 12 kilometers above Earth from an aircraft-borne infrared telescope (the Kuyper Airborne Observatory), mapped the far-infrared emission from warm dust in the central region with the highest resolution up to that time (30 arc seconds) These observations revealed that the inner 5 or 6 light years were relatively empty of interstellar dust and gas. However, between 6 and 15 light years there appeared to be a ring of warm dust. In 1989 Harvey Liszt, Butler Burton, and Thijs van der Hulst found, via 21-cm neutral hydrogen line observations at the VLA, that this ring consists largely of neutral gas, that it is rotating about the center with a velocity of about 100 km/s, and that it is somewhat inclined to the plane of the Galaxy as defined locally.

Since then, the ring has also been detected in the spectral lines of a number of molecular species such as carbon monoxide and hydrogen cyanide, and these lines help to define its properties and motions. The gas distribution in the ring is clumpy with densities on the order of 100 000 particles per cubic centimeter, and the total mass of gas is on the order of 10 000 solar masses. This "circumnuclear disk" or CND is similar to the torus found in the central light year of NGC 4258. Such dusty tori surrounding central black holes may well be a general feature of more active galaxies. As suggested by Peter Barthel in 1989, then at Caltech, this may account for the distinction between quasars and radio galaxies – radio galaxies being those objects in which the view to the central object is obscured by the dusty disk, whereas quasars are seen when viewing the object from a more face-on position where the central source is unobscured.

The ionized gas filaments, evident as the mini-spiral, are found within the inner boundary of this ring, but the total mass of gas in the ionized filaments is much smaller – perhaps only 100 solar masses. The spatial relationship between the filaments and the CND is shown in Figure 9.3. The CND is shown here in the intensity of the 3.3-millimeter line of hydrogen cyanide from observations made at the Hat Creek (California) interferometer by Rolf Güsten (Max Planck Institute for Radio Astronomy) and his colleagues. From this figure it becomes evident that at least part of the "spiral structure" observed in the filaments is actually due to

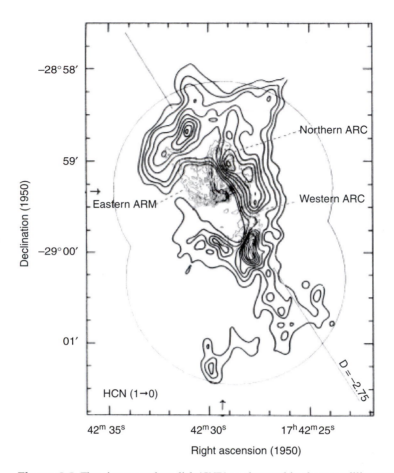

Figure 9.3 The circumnuclear disk (CND) as observed in the 3.3-millimeter line of hydrogen cyanide by Güsten et al. (1987). The faint inner contours show the radio continuum emission from the ionized filaments, the mini-spiral. From such observation it became clear that at least part of the mini-spiral (the "western arc") was actually the inner ionized edge of the CND. The diagonal faint straight line is parallel to the Galactic plane defined locally.

the CND, particularly the so-called "western arc" which appears to be the ionized inner edge of the CND.

If the CND is in pure rotation about the center of the Galaxy, then that would provide a limit on the mass within inner 10 light years, and this turns out to be somewhat less that 10 million solar masses. This, give or take 30% or 40%, is about what would be expected from the old stars distributed about the center and so places no significant constraint on the mass of an additional component – a black hole – of a few million solar masses. But what about the velocity of the ionized filaments further in? As mentioned in Chapter 6 there is a spectral line in the mid-infrared, the 12.8-micron line of singly ionized neon (the neon atom with one electron removed), and this line is emitted by denser clumps in the ionized filaments. Moreover, the velocity indicated by Doppler shift of the line varies systematically with distance from Sgr A*, particularly along the so-called "northern arm" in the mini-spiral.

In 1985 Eric Serabyn and John Lacy (University of California at Berkely) modeled the systematic variation in the velocity of the neon-emitting regions along the northern arm by motion along a parabolic orbit in a gravitational field produced by an extended distribution of stars plus a point mass in the center. The basic idea behind such a model is that a small infalling cloud would be tidally stretched along its orbit and so would actually trace the orbit. Serabyn and Lacy found that the results were most consistent with a point mass of three to four million solar masses. But the principal caveat is that gas is subject to various nongravitational forces – hydrodynamical disturbances such as shocks from explosions, radiation pressure, and winds from young stars. Oort in his review in 1983 even argued that the ionized gas streamers could be jets from the central object, that is, outflow instead of inflow. So in spite of the success of the orbit model, there is no guarantee that we are really observing motion in a gravitational field. It would be useful to have another tracer – a test particle tracer.

9.4   Distribution and Motion of Young Stars Near the Galactic Center

The proposal by Rieke and Lebofsky that the Galactic Center could be powered by massive young stars rather than a single point-like source at the very center was corroborated by David Allen and his collaborators when they realized that many of the bright infrared sources near Sgr A* are in fact the young hot stars with a total power capable of explaining the total far-infrared luminosity of the central ten light years (re-radiation of the star light by warm dust particles). These hot stars are also responsible for the ionization of the inner edge of the neutral dust ring and the spiral-like filaments. Allen and collaborators found the blue stars that accompanied the Rieke and Lebofsky starburst inferred by the presence of evolved red supergiants.

Then, in 1990, it was discovered that many of these stars emitted a spectral line – an emission line from neutral helium at 2 microns. These could only be young blue supergiant stars that are losing mass through winds (driven by radiation pressure) at a high rate, up to 10^{-4} solar masses per year. This may not seem like much, but it means that one solar mass is lost per star in 10 000 years. They clearly must be very young stars. The group of Reinhard Genzel (Max Planck Institute for Extraterrestrial Physics, see Figure 9.4), using their own advanced imaging spectrometer at the Anglo-Australian Telescope (AAT), followed soon by

Figure 9.4 Reinhard Genzel of the Max Planck Institute for Extraterrestrial Physics has pursued, and caught, the massive black hole in the Galactic Center. (Courtesy of ESO/Max Planck Institute for Extraterrestrial Physics.)

Michael Burton and David Allen using the new infrared array camera of the Anglo-Australian Observatory, were able to map a number of these helium emission line stars near the Center. They form a compact cluster in the inner few light years.

The important aspect of emission line stars is that astronomers can measure their velocity along the line of sight using their Doppler shifts; these stars become dynamical probes of the gravitational field – probes that are not sensitive to the various nongravitational forces that can act on the gas. This aspect was quickly exploited by Genzel's group, who proceeded to measure the velocity of these objects using the AAT and the recently developed NTT (new technology telescope) of the European Southern Observatory in Chile. Their conclusions were summarized in 1996 when they reported the results of the analysis of the measured velocities of 223 stars on scales measured down to one-third of a light year of Sgr A*.

Although the stars near the center may be slowly rotating systematically about the center as does the gas in the CND, they are primarily moving randomly, in all directions. This random motion supports the system of stars against gravity; it is like a gas held up against gravity by pressure, but instead of molecules the gas particles are individual stars. The analysis for such a star gas, the determination of the mass distribution, is carried out by using the Jeans equation, named for James Jeans, the British physicist who first applied this equation to a system of stars. It is basically an equation of hydrostatic equilibrium that equates at every point the pressure force of the star gas pushing outward to the gravitational force pulling inward. Figure 9.5 illustrates the Jeans equation result by Genzel and collaborators

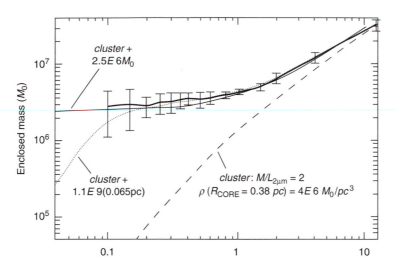

Figure 9.5 The mass enclosed (in solar masses) within a given radius (in units of parsecs which is about 3 light years) as implied by the line-of-sight velocities of the helium emission line stars analyzed via the Jeans equation (Genzel et al. 1996). The dashed line is the mass distribution resulting from the old stars alone, assuming that the density distribution levels off at about 1 light year, and the dotted line shows the mass distribution assuming the distribution of stars with a central density of thirty million solar masses per cubic light year within 0.3 light year. The light solid line shows the implied mass distribution with the distribution of stars plus a 2.5-million solar mass point mass. The points with error bars are from the observed distribution of velocities as analyzed via the Jeans equation.

for the mass distribution; this is a plot of the mass enclosed within a given radius as a function of that radius. For a cluster of stars, that mass should decrease to zero at zero radius. But for a cluster with a point mass, it should level off at the value of that point mass. This is precisely what it appears to do at around 2.5 million solar masses.

But there is also a caveat here. Unlike for a normal gas, where the spread in the velocity of particles is the same in all directions (the velocity field is isotropic), for a star gas this is not necessarily the case. For the gas of stars the spread in velocities can be different in different directions, and this is an ambiguity for an analysis based on the Jeans equation. When only one component of the velocity is measured, then there is no way of knowing whether or not the velocity and hence the pressure is isotropic, and this can make a very big difference for the estimated mass distribution. But the problem can be solved if all components of the velocity of the stars can be measured.

The line-of-sight velocities of the young stars near the center range up to 400 km/s. If the velocities of the stars perpendicular to the line of sight are comparable, then we should able to see the stars move across the sky by up to 50 milli arc seconds in 4 years (this is called proper motion). With the imaging capabilities of large telescopes developing rapidly in the 1990s (discussed in Chapter 10) it was becoming possible to measure this perpendicular motion of stars at the Galactic Center. The first measurement of proper motions of thirty-nine stars observed to be between 1/100 and 1/10 of a light year of Sgr A* were reported by Andreas Eckart and Reinhard Genzel in 1996. The inferred velocities perpendicular to the line of sight were consistent with the spectroscopically determined velocities parallel to the line of sight. The velocity spread was equal in all directions; the velocity distribution was isotropic. This appeared to prove the validity of the Jeans equation calculation; it really did seem that there is to a mass concentration of two to three million solar masses in the inner few hundredths of a light year.

But then this brings us back to the original problem. How can stars form so close to such a mass concentration in view of the strong tides? This is the paradox of youth, a phrase coined by Mark Morris and first aired publically in 2003 by Andrea Ghez and her colleagues at UCLA. It stuck.

9.5   Star Formation in the Near Tidal Field of a Black Hole

There are several ideas about how such young stars could be found in the near vicinity (within one or two light years) of a black hole with a mass of several million solar masses. A trivial possibility is that this is a chance superposition – the stars are not actually near the Sgr A* in three-dimensional space but lie along the line of sight, many light years in front of or behind the black hole. But it would be indeed remarkable if such a concentration of stars with their apparent centroid so near Sgr A* were just such an unlikely projection, and the fact that the spread in velocity increases nearer the position of Sgr A* would certainly seem to rule out this explanation. Another idea is that the stars were actually not formed near the black hole but have migrated inward from further out. But these stars are very young – less than a few million years – the migration must be very rapid indeed (only a few orbital periods). A third possibility is that the young stars are formed *in situ*, where they are seen within 1 light year of the black hole. But then the density of gas must have much higher in the recent past than it is observed to be at present in order to overcome the tidal constraint.

In his 1988 lecture Phinney reminded us that it is possible for high gas densities to be generated in strong shock waves resulting from collisions between gas streams with relative speeds in excess of 100 km/s. Such strong shocks could develop when gas clouds fall toward the center, wrap around the black hole, and self collide. I carried this idea further with a simulation of such a small gas cloud

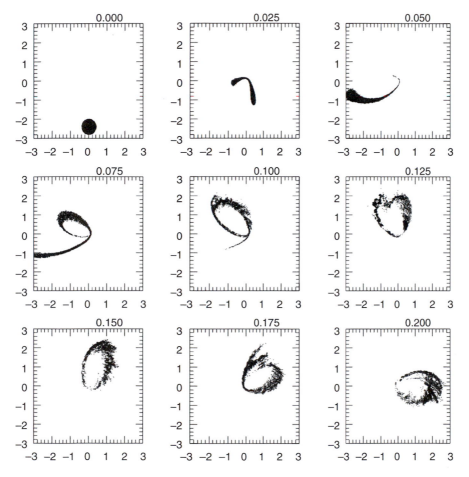

Figure 9.6 A simulation of a small cloud falling into a gravitational field resembling the Galactic Center, that is, with an extended star cluster and a central point mass of 2.5 million solar masses. Each frame is a snapshot of the gas distribution every 25 thousand years; the distance units on the X and Y axes are in parsecs (about three light years). The cloud wraps around the point mass, self-collides with resulting strong shocks, which may lead to the formation of stars. The structure continues to precess keeping the same shape for about one million years (Reproduced from Sanders 1998).

moving toward but just avoiding a 2.5 million solar mass object at the center of a gravitational potential resembling that of the Galactic Center.

The results are shown in Figure 9.6 which, is a time sequence in units of 1 million years; each frame is a snapshot of the cloud at every 25 thousand years. The x- and y-axes show the distance from the putative black hole in units of parsecs (about 3 light years). The simulation is two-dimensional (in the orbital plane of the cloud), but this seems fair because a spherical cloud will collapse rapidly into the orbital plane to form a disk. We see that the cloud has wrapped around the black hole in frame 3 (at about fifty years after the beginning of the infall) and self-collides in frame 4 (100 000 years). It then forms an elliptical ring that maintains its shape as it precesses around the center for up to 1 million years. Precession means that the gas is flowing along the pattern much faster than the pattern is moving, and in this case the pattern precesses in the opposite sense of the motion of the particles. It is in the self-collision that strong shocks form and, presumably, stars. Not entirely coincidentally, the velocity along the precessing filament matches that of the northern arm of the mini-spiral quite precisely.

Such a model has several positive aspects. As in the observed northern arm the gas wraps around, but does not contain the black hole (Sgr A*). Because the elliptical ring maintains the same shape as it precesses around the center, it is not such a short-lived apparition; the structure can be seen for up to 1 million years instead of the initial orbit time of 100 thousand years (and so addresses the objection of Oort against the in-falling cloud model). Because the cloud collapses into a disk, the stars formed will also be found in a disk in the original orbital plane, which, because of the random nature of the cloud infall event, can have any relation to the Galactic plane. The formed stars will be found on highly elliptical orbits, rather than circular orbits, which is another observed characteristic of the ensemble of young stars.

What could cause a cloud to suddenly plunge toward the center? The interstellar medium in the inner 10 to 50 light years is observed to be quite turbulent; that is to say, the gas is distributed in clouds of various sizes with a large component of random motion in addition to rotational motion about the center. This random motion could be maintained by the occasional flaring of the black hole or the occasional starburst like that we are presently observing. Occasionally two clouds moving in opposite directions would collide, losing random energy and orbital angular momentum and then fall rapidly toward the center. Perhaps a cloud like that simulated here arose in the turbulent medium in the circumnuclear ring. This could be the mechanism of both star bursts and sporadic flaring of the black hole.

The black hole appears to be necessary for these high velocities and strong shocks. So the black hole, rather than suppressing star formation, may actually promote it. But there are other models for the formation of young stars near the black hole, and I discuss these in Chapter 10 after I describe the clear observational proof for a massive central object – an object that can plausibly only be a black hole.

10

Stellar Orbits in the Galactic Center, QED

10.1 Two Mountains, Two Legends

The Hawaiian islands in the mid-Pacific are quite young on geological timescales; in fact they are still forming. The volcanic island chain runs from northwest to southeast; as the Pacific plate drifts northwestward over a weak spot in the underlying mantel, magma, hot molten basalt, occasionally bubbles up to the seabed and onto the surface forming a new island. The most recent consequence of this process is the big island of Hawaii which is less than 400 thousand years old. Its principal volcano, Mauna Kea, stands at 4207 meters (13 803 feet) above the sea, but measured from its base at the bottom of the sea it is 10 200 meters (33 500 feet), taller than Mount Everest at 8848 meters. It is a significant mountain. Although the climate of the island is wet, the summit of Mauna Kea stands above moisture in the atmosphere. It is above 90% of the water vapor and above 40% of the atmosphere itself. This makes the site ideal for astronomical observations, and since the 1950s when the first solar observatories were constructed, there has been continuing development of the summit for astronomy.

Mauna Kea is sacred for the indigenous Hawaiian people; the ancient belief is that many of their principal deities live at the summit, which plays a role rather like Mt. Olympus in Greek mythology. In the past, major chieftains ascended to the top to commune with the deities. Continuing reverence for the mountain has made the issue of observatory development quite sensitive.

Interestingly, there is an ancient Hawaiian legend that has some relevance to the formation of the island chain and the mountain. In this story, the big island is ruled by Poliahu, the goddess of snow. But one day, the goddess of fire, Pele, also appears (apparently in the Hawaiian pantheon these are more in the nature of demigods rather than true gods). Now Pele, having this terrible power at her disposal, was not very popular so she was chased from island to island, from northwest to southeast, until she finally arrived at the big island (this presages the scientific narrative of the volcanic history of the archipelago). When Pele arrived on Hawaii she was not immediately recognized, and challenged Poliahu to a sledding contest. It was close, but three other goddesses judged that Poliahu was the winner (a sort of judgment of Paris moved to a Pacific setting). Pele flew into a rage and created an eruption of fire and lava on Mauna Kea; Poliahu fled to the summit and extinguished the fire with a covering of snow, and it has been so ever since. But the story does imply a continuing struggle between extended periods of peace and quiet punctuated by shorter epochs of violent eruption.

The second mountain is in the Atacama desert of Chile, the driest region of the Earth. The cold Humboldt current offshore in the Pacific causes an inversion layer, so the rain that might fall on the land falls on the ocean instead. And to the east the region is shielded from the moist trade winds of the Amazon basin by the principal range of the Andes. The landscape is lunar, or perhaps Martian because there is, at times, moisture bearing fog providing lower lying regions with sparse vegetation. There were once lakes but now these are composed primarily of salt and sand. The desert is rich in minerals, such as sodium nitrate and copper and since the nineteenth century has been a center of major mining activity.

Cerro Paranal is located about 120 kilometers south of the coastal town of Antofagasta and about 12 kilometers inland. With an altitude of 2635 meters it is not so high as Mauna Kea, but its extreme dryness and relative isolation from human sources of light and radio emission make it also an ideal site for astronomical observations. There are no protests over the sanctity of the mountain because the density of the human population is, and has been since pre-Columbian times, very sparse in this inhospitable environment.

There is a legend of the desert, however – a legend that is actually quite relevant to the story here. It involves a mythological creature and is certainly not pre-Columbian but rather from the advent of mining activity (although its roots may be pre-Columbian). The creature is a bird, the *Alicanto*, that sits quietly in a cave in the mountains of the desert but occasionally feeds on gold or silver. When it does it shines brightly with the colors of those elements – so brightly that it can cast shadows at night – and strange lights, like stars, emanate from its eyes. The miner who spies this creature and follows it may find its food and become fabulously wealthy. But, if the Alicanto detects that it is being followed it will lead the miner to a deep shaft where he will fall and perish.

10.2   New Technology for a New Millennium

In the mid-1980s the W.M. Keck Foundation donated US 70 million for the construction of the first of two enormous telescopes on the (hopefully) extinct volcano of Mauna Kea. The new telescope was to be 10 meters in diameter and consist of thirty-six hexagonal independent elements controlled by computer to function as a single mirror; the precise position of the hexagonal mirrors was adjusted continually to maintain the correct shape. The first telescope, completed in 1993, was to be followed by a second (1996), and the two could work together as an interferometer (Figure 10.1).

At about the same time, the governing council of the European Southern Observatory, with representatives from all of the member countries, decided to construct the Very Large Telescope, the VLT, which would consist of four 8.2 meter telescopes that could, in principle, also be used as an interferometer (Figure 10.2). The mirror of each of these telescopes is the more traditional single unit, but it is deformable. The mirror rests upon by 150 axial actuators, computer-controlled supports, that apply pressure and modify the mirror's shape to the optimum form as the telescope tracks across the sky. The first of the VLT telescopes came on line somewhat after Keck, in 1998, and the final telescope in 2000. These were constructed on the mountain of Paranal in the Atacama desert of northern Chile.

In both cases, the Keck telescopes and the VLT, the "active optics," the continual adjustment of the shape of the mirror due to stresses incurred during tracking, is made possible by high-speed computers; these large telescopes are, in

Figure 10.1 The Keck telescopes near the summit of Mauna Kea on the big island of Hawaii. The elevation of 4145 meters, the large aperture of the primary mirrors, and the laser guide star adaptive optics system make this an ideal instrument for mapping positions and motions of stars near the Galactic Center. (Photo from NASA/JPL.)

Figure 10.2 The Very Large Telescope (VLT) of the European Southern Observatory on Cerro Paranal (altitude 2635 meters) in Chile. Each of the four primary mirrors has a diameter of 8.2 meters with laser guide star adaptive optics for infrared observations on one telescope. This advanced system, the high altitude, and very dry climate provide an ideal observatory for observations of the Galactic Center stars. (Figure from ESO.)

that sense, a direct consequence of the ongoing exponential growth in computing power known as "Moore's law."

Why are such large telescopes useful? What can they do that smaller aperture telescopes, such as the Hubble Space Telescope, cannot do? First of all, the larger surface area means more light collecting – fainter objects can be seen. But second, and more importantly, there is more resolving power – the ability to separate close-by images on the sky, to see finer detail. The resolving power of a telescope is measured in terms of the minimum angular separation of two point sources that can be distinguished as two separate sources. For example, a telescope that can distinguish two stars separated by one arc second has a resolving power twice as large as a telescope that can only distinguish sources separated by two arc seconds.

There is a theoretical maximum to the resolving power that is inversely proportional to the size of the primary mirror (or lens) of the telescope but directly proportional to the wavelength of the radiation being observed: it is called the diffraction limit and is given by

$$\theta \, (\text{arcsec}) = \frac{\lambda \, (\text{microns})}{4D \, (\text{meters})}$$

where λ is the wavelength in microns and D is the diameter of the mirror in meters. So the 10-meter mirror of a Keck telescope can resolve sources separated by 0.05 arc seconds at a near-infrared wavelength of 2 microns.

This would be the maximum resolving power of a 10 meter telescope; in practice, the resolving power is lower (unless the telescope is in space) primarily because of the turbulent atmosphere of the Earth. The atmosphere, from top to bottom, consists of small elements with different speeds and slightly different densities; it is like a boiling pot of water with different parts in relative motion to one another. Each of these elements acts as a separate lens and distorts a beam of light coming in from space in different ways. The overall effect, over time, is to blur the image and degrade the resolution so that, under most circumstances, the resolving power is reduced to about 1 arc second under good conditions – far from the theoretical maximum of the 10-meter telescope on the surface of Earth. This blurring is traditionally called "astronomical seeing."

But, there are different techniques for overcoming this blurring effect of the atmosphere. One obvious technique is to move the telescope to space, as in the case of the Hubble Space Telescope, but this is prohibitively expensive for a very large telescope. A ground-based technique is called "speckle imaging," in which many short-exposure pictures are taken of a particular source, so short that for any one image the optical path of the light through the atmosphere does not vary due to the turbulent motion; such an image would have the original maximum resolution for the telescope given by the little formula above. But because the exposure time is so short, the image is not very bright – not many photons are actually collected. So the trick is to add up or stack this series of images by digital processing later on – post-processing. For visual wavelengths this is difficult because the maximum exposure time of one image must be very short – on the order of 10 milliseconds (1/100 second). That is the coherence time of the atmosphere to visual light; for time intervals longer than the coherence time, the multiple images of a star – the speckles – caused by different turbulent elements along the line of sight through the atmosphere move significantly, causing the blurring. It is easier to do this at infrared wavelengths because the coherence time is ten times longer at 100 milliseconds. In the infrared more photons can be captured in a single exposure and fainter objects can be observed. Nonetheless the method is not very effective for faint images.

A more powerful technique is "adaptive optics," which corrects the optical system in real time for the distortions in the incoming waves of electromagnetic radiation. This is done over a time interval that is short compared to the coherence time of turbulence in the atmosphere, that is to say, many times per second. Adaptive optics should not be confused with "active optics" mentioned earlier; active optics corrects the shape of the primary mirror on a much longer timescale – the time it takes for the telescope to track an object across the sky as the Earth rotates – on the order of hours. With adaptive optics very rapid corrections (every few milliseconds) are made to the shape of an additional mirror along the optical path to correct for the atmospheric turbulence.

What does the turbulence do to light or infrared radiation? Basically the light entering the Earth's atmosphere from a distant astronomical source can be considered as plane parallel waves; that is to say, the crests of the waves may be represented as parallel lines perpendicular to the direction of the radiation. But as a result of passing through the turbulent atmosphere, the wave crests are no longer parallel but distorted; for astronomical observations with large telescopes, the most troubling distortions are caused by atmospheric turbulent elements with a size on the order of one-half meter, smaller than the primary mirrors so there are a number of the fluctuating distortions, light and dark patches, over the area of the mirror. In the adaptive optics system there is a wavefront detector that samples part of the incoming light, determines the shape of the wavefronts, sends a signal to the computer that calculates corrections, which then sends these on to attenuators on the deformable mirror (not the primary mirror, which cannot respond so quickly). These little pistons alter the shape of the mirror to correct the wavefronts again to the parallel form they had before entering the atmosphere (see Figure 10.3 for a simplified description of this technique).

Because turbulence is different at different depths and directions in the atmosphere, the wavefront detector needs to sample light coming almost along the same path through the atmosphere as the object being observed, the source, which itself is usually too faint for effective determination of the form of wavefront. So, the astronomer must look for a bright guide star in the near vicinity of the source. The problem is that most often there is no bright guide star available near any given faint source in a small field of view, so the solution is to make an artificial star high in the atmosphere – an artificial star using a laser.

Ninety-five kilometers above the surface of the Earth there is a layer of sodium atoms created by the breakup of micro-meteors in the upper atmosphere. A laser tuned to a wavelength of 5890 angstroms, the wavelength of the principal sodium line, excites this transition; when the sodium atoms decay back to the ground state, the result is a yellow spot in the sky, an artificial guide star that can be placed near any source (see Figure 10.4).

Adaptive optics has a history stretching back for half a century. It began in 1953 when American astronomer Horace Babcock wrote a paper on the "possibility of compensating astronomical seeing." This laid out the basic principle of correcting the distorted wavefronts by deforming a mirror along the optical path. Because of the technical difficulties not much was done until the early 1970s, when the U.S. military became interested in correcting the effects of atmospheric blurring in identifying Soviet satellites, but this work remained classified until 1992 after the end of the Cold War. There was, however, independent work by astronomers;

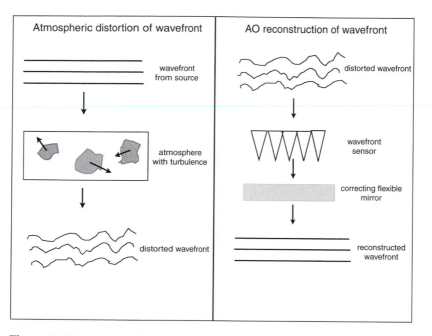

Figure 10.3 Basic principles of adaptive optics. The left-hand panel shows the incoming wavefront being distorted by turbulence in the atmosphere. The right-hand panel shows reconstruction of the distorted wavefront by the adaptive optics system. Here the shape of the wavefront (generated by a natural or artificial laser guide star) is detected by a sensor, and corrections are calculated by computer and sent to attenuators that change the shape of the correcting mirror appropriately. Ideally, the original incoming wavefront (before encountering the atmosphere) is reconstructed before being sent on to the camera.

in 1975 two French scientists, Reynaud Foy and Antoine Labeyrie, published a short ground-breaking paper in the European journal *Astronomy and Astrophysics* on the "feasibility of an adaptive telescope with a laser probe." This was the first published description of the possible use of the sodium laser in generating an artificial guide star. With the declassification of military work on adaptive optics in the early 1990s, an active collaboration was formed between Lawrence Livermore Laboratory and the University of California to provide an adaptive optics system on the Keck telescopes.

The advantages of adaptive optics are obvious in Figure 10.5, which is taken from a paper of Ghez et al. in 2008. The left panel shows the central 8 arc seconds (1 light year at the Galactic Center) image obtained in 2008. The panels on the right are blowups of the inner arc second (1.5 light months) where the upper image is that obtained by Laser Guide Star Adaptive Optics, (LGSAO), while

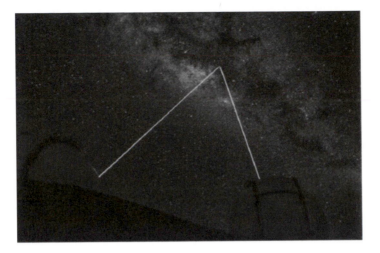

Figure 10.4 The yellow light laser, in this case pictured at the Keck telescopes, excites a transition in atomic sodium at an altitude of 95 kilometers. This produces an artificial guide star at any location on the sky for measuring the wavefront distorted by the turbulent atmosphere. The active optics system then corrects the wavefront from the nearby astronomical source of interest. (Image by Elton Tweedie Photography, courtesy of W.M. Keck Observatory.)

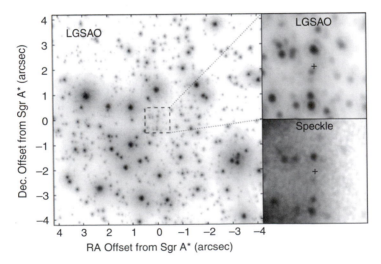

Figure 10.5 The power of adaptive optics. The panel on the left shows an adaptive optics image of the inner 4 seconds of arc of the center of the Galaxy made with the Keck telescope. In the left the upper panel is a blow up of the inner one second of arc while below is the same region imaged via the speckle technique. Far more stars are visible in the AO image. From Ghez et al. (2008), with permission of Andrea Ghez.

the lower image is that obtained by the speckle technique. The improvement is evident as far more stars are seen to emerge from the background noise.

We see that adaptive optics is the ideal tool for infrared observations of fields that are very crowded with faint objects. This is particularly true of the observatories on Mauna Kea and Paranal, which are high and dry, but also, being above a cold ocean, experience less ground level turbulence. The large apertures have enormous light gathering power and, with adaptive optics, resolutions of 50 milli-arc seconds. The positional accuracy is even higher – down to 10 milli-arc seconds. In other words, an ideal project for these observatories is precision imaging of the stars near the Galactic Center, where 50 milli-arc seconds corresponds to 200 000 light seconds or about 60 billion kilometers. This is about four thousand astronomical units, or about 5000 Schwarzschild radii for a four million solar mass black hole. These advantages were realized by two groups, one European and the other American, who were eager to exploit these new facilities in order to trace the paths of stars very near Sgr A*, the putative black hole at the Galactic Center.

10.3   Inward Bound: Leapfrog to the Black Hole

At the turn of the millennium a neck and neck race began between groups at the Keck and the VLT (the US vs. the EU). The goal was to measure the proper motions over time intervals of several years and, thus, not only the velocities but also the actual orbits of individual stars.

The European group was led by Reinhard Genzel of the Max Planck Institute for Extraterrestrial Physics near Munich. Genzel had been pursuing this problem for twenty years, and, using speckle imaging on the 3.5-meter New Technology Telescope of ESO, had observed the motions of stars in the central light year. These observations already gave convincing evidence of a highly concentrated mass distribution near Sgr A*.

The American group was led by a relative newcomer in this field: Andrea Ghez of the University of California at Los Angeles (Figure 10.6). Ghez received her Ph.D. in 1992 at Caltech where she had specialized in observations going down to the theoretical resolution limit of the telescope – techniques such as speckle imaging. When she went to work at UCLA, she was, in a real sense, the right woman at the right place at the right time. There was an existing strong group at UCLA with an interest in the Galactic Center, and with Ghez the group was quick to take advantage of the possibilities offered by the Keck telescope.

As we saw in Chapter 9, many of the relatively bright young stars near Sgr A* exhibit the strong helium emission line. This allows astronomers to measure their radial velocity, the component of velocity along the line of sight. In 1995 Genzel

Figure 10.6 Andrea Ghez uses the adaptive optics system on the Keck telescope to trace the orbits of stars very near the center of the Galaxy. In this way she and her group at UCLA, along with that of Reinhard Genzel at the Max Planck Institute near Munich, have placed the tightest constraints on the mass of the black hole and ruled out suggested alternatives. Courtesy of Andrea Ghez.

and his colleagues carried out a systematic survey of about two dozen such stars at distances of 1 to 10 arc seconds (1/8 to 1 light year) from Sgr A* and found that the radial velocity increased toward the center; this constituted evidence (via the Jeans equation) for a mass concentration within one and one-half light months.

Then in 1996 Andreas Eckart and Genzel published proper motions for thirty-nine stars with a time baseline of four years. These stars were observed down to a projected distance of one-tenth of 1 light year from the Galactic Center, and the velocities perpendicular to the line of sight were comparable to the radial velocities, apparently validating the Jeans equation approach. The implied mass within 0.1 light year was two and one-half million solar masses.

Between 1995 and 1997, Ghez and her colleagues had been observing the central few arc seconds of the Galaxy using the Keck – the first Galactic Center observations making use of the new 10-meter telescope with image sharpening, in this case speckle imaging allowing resolution down to the 50 milli-arc second limit of the telescope for the brighter stars. In 1998 they published their results on proper motions of ninety stars within 4 arc seconds of Sgr A*. The number of stars with measured proper motions was doubled, and, in the inner 1 arc second, the accuracy of the measurements was improved by a factor of four. The group began their own nomenclature system for these stars lying near Sgr A* – the "S" stars, ranked in order of their distance from the compact radio source, designated by two numerals: S0-2, for example, would be the second star (designated arbitrarily) from zero to one arc second, S1–5 would be the fifth star between 1 and 2 arc seconds, etc. (the MPE group had a separate system, simply ordering the S stars in no systematic fashion, i.e., S1, S2, ..., S95). The highest proper motion observed at the Keck implied a velocity, perpendicular to the line of sight, of more than 1500 km/s – certainly the highest stellar velocity ever measured in the Galaxy up to that point. The Ghez group saw the increase in the spread in velocity with decreasing distance from Sgr A*, as had been observed earlier by the Genzel group. In fact, the results were completely consistent with those of Genzel and colleagues, implying the presence of a mass of two and one-half million times that of the Sun within a few tenths of an arc second.

But a completely new result followed in the year 2000. The UCLA group was able to measure not only the velocities but also the accelerations of the stars, the change in velocity over time. Acceleration, like velocity, is a vector – it has direction as well as magnitude – and Ghez and her collaborators found that the accelerations vectors pointed directly back to Sgr A* as the location of the center of force. This was an extremely important observation because it had always been supposed that Sgr A* marked the location of the massive object in the center. Here was the first actual direct proof.

In 2000 Genzel's group, using the New Technology Telescope of ESO (3.5 meters), followed up their initial work on proper motions of the bright stars near the center. For a number of the young stars they found significant deviations from a purely random velocity pattern; the stars seemed not to be moving in all directions but exhibited a clockwise pattern of circular motion about the center. Then, in 2003, Yuri Levin and Andre Beloborodov (Toronto), using these same data, found that many of these stars appeared to lie in a single plane; they comprised a disk with a significant component of rotational motion. These are young stars, so Levin and Beloborodov proposed that they were formed relatively recently (one or two million years ago) in a dense gaseous accretion disk around the black hole (dense to get around the tidal limit on gravitational collapse imposed by the massive black hole). The gas disk was subsequently accreted by the hole, or ejected in the star formation/accretion event, leaving only the remnant of stars. In 2003 and 2006, based on extended observations, this disk was confirmed by Genzel's group who presented evidence for a second disk of stars in a different plane. This has consequences for the origin of young stars in the center as well as the accretion onto the black hole.

10.4   Individual Star Orbits: The Black Hole Gains Weight

By the year 2000 it had became possible to trace the individual orbits of stars very near the Galactic Center, at both the VLT and the Keck. This was due primarily to image sharpening techniques on these new 10-meter class telescopes.

Speckle imaging was applied at first at both facilities, but in 2002, the VLT was equipped with an infrared wavefront detector. This allowed the European group to make use of a natural infrared guide star (IRS 7, the bright point source originally found by Becklin and Neugebauer), and so use adaptive optics for the Galactic Center sources – for the first time on a 10-meter class telescope. Two years later, in 2004, the artificial laser guide star (LGSAO) was implemented at Keck, and the AO image sharpening technique was then available to the American group as well.

The use of individual orbits to determine the gravitational field and hence the mass distribution is to be greatly preferred over the Jeans equation – the equation of hydrostatic equilibrium for a system of stars. To use the Jeans equation one must measure the spread in velocity at a given radius – the velocity dispersion. To precisely define the velocity dispersion, a large number of stars is required; otherwise, some stars with higher velocity will be missed and the velocity dispersion will be underestimated. This leads to an underestimate of the mass. So there is a bias in the use of the Jeans equation for a very finite number of objects – a bias toward mass estimates lower than the true mass. With individual orbits, this bias disappears.

The true shape of orbits in the potential of a point mass is known from the time of Kepler: the orbits are ellipses with the center of attraction at one focus of the ellipse (an ellipse has two points, the foci, such that the sum of the distances from these foci to any point on the ellipse is constant). Ellipses in three dimensions project into ellipses on the plane of the sky, but the difference in position of an apparent focus from that of Sgr A* can help in sorting out the inclination and orientation of the true ellipse. If the mass distribution deviates significantly from that of a point mass, then elliptical orbits will precess; for a perfect point mass the ellipses are fixed in space.

In their work published in 2000 presenting the accelerations of three stars within 0.02 light years of Sgr A*, Ghez and co-workers were able to fit orbits to the individual stars. The star S0–2 in particular had two orbital solutions with periods of 15 and 550 years. The ambiguity was due to the relatively short time base of five years; it was just not possible to find unique orbit over such a small fraction of a period. But then, in 2002, the Genzel group, with a time base of ten years, was able to restrict the orbit of S0–2 to the shorter period alternative (fifteen years), and presented the first publication of a nearly complete orbit (Schödel et al. 2002). The best fit for the black hole mass was increased to 3.6 million solar masses.

This was followed in short order (Ghez et al. 2003) by the UCLA group who had detected absorption lines in the spectrum of S0–2 (hydrogen and helium lines) and thereby increased the precision of the orbital determination by a factor of two to three. The estimated period of the orbit was increased slightly to about sixteen

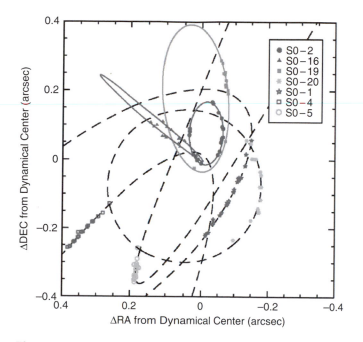

Figure 10.7 Orbits of seven stars in the inner light month obtained at the Keck telescope using primarily speckle imaging. Elliptical orbits projected onto the sky are also ellipses. (From Ghez et al. 2005 courtesy of A. Ghez.)

years, and the best value for the black hole mass was increased to 4.1 million solar masses.

The two groups appeared to be spurring each other on to mapping more orbits with higher precision. Six months later the MPE group published the orbits of six stars with a baseline of ten years, making use of data taken in the beginning with speckle imaging on ESO's 3.5-meter New Technology Telescope and finally with the AO system on the VLT (Schödel et al. 2003). Now the preferred value of the mass of the black hole had slipped back a bit to 3.4 million solar masses.

The first publication of seven individual orbits by the Ghez group appeared in 2005 using AO to track the stars after 2004. This was followed again six months later by a similar study of seven orbits was published by the Genzel group (Eisenhauer et al. 2005). The results of the two groups are shown in Figures 10.7 and 10.8, where we see a striking similarity especially for the three closest orbits. That there is such agreement between two competing groups using different instrumentation on different telescopes provides a strong confidence in the results. In both studies the black hole mass remains at 3.6 to 3.7 million solar masses.

The next improvement in the quality of the analysis, now using the full power of adaptive optics on both telescopes, followed almost simultaneously in

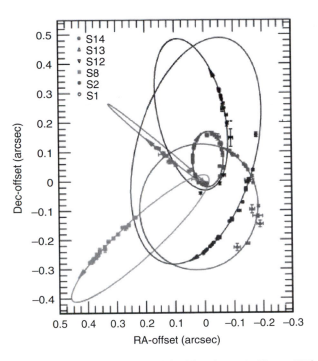

Figure 10.8 Six of the star orbits (also shown in Figure 10.7) obtained at the VLT using adaptive optics after 2002 with the natural guide star. Note the similarity of the results with the UCLA group for the three orbits closest to Sgr A* S0-2 (S2), S0-19 (S12) and S0-16 (S14). From Eisenhauer et al. 2005, courtesy of R. Genzel.

2008/2009. Both groups claimed very high positional accuracy, 0.2 to 0.3 milli-arc seconds (about two light hours at the distance of the Galactic Center). The UCLA group focused on the closest, shortest period star, S0-2, which best defines the inner mass distribution, and the MPE group presented precision orbits of twenty-eight stars.

Assuming Keplerian orbits, the observed paths – the position on the sky along with the three-dimensional velocities – can be matched by a model in which the adjustable parameters are the mass of the central object and the distance to the Galactic Center. In these more precise results from 2008, the central mass turns out to be between four and four and one-half million solar masses (4.0×10^6 – 4.5×10^6 $M_\odot$) and the distance is between 26 000 and 27 000 light years. The implied distance to the center is essentially the same as the distance estimates determined by other methods, but the black hole has apparently gained some weight. This is due to the bias in estimating the mass using Jean's equation – the fact that the velocity dispersion, the total spread in velocity, is underestimated with a relatively small number of stars measured.

The orbital period of S0–2, the time it takes to complete one cycle about the black hole, is only sixteen years (it has now been tracked for one complete period), and it passes within 0.0017 light year or about 15 light hours of the black hole; this is 1400 times the Schwarzschild radius of a four million solar mass black hole. The precision with which this orbit has been determined and the fact that there is no measurable deviation from a point mass potential means that the radius of the central object must be less than the point of closest approach. This then implies that the density of the central object must exceed 2×10^{14} solar mass stars per cubic light year, equivalent to 3.5×10^{15} hydrogen atoms per cubic centimeter.

This limit on the radius and the minimum density of the central object essentially eliminates all conceivable alternatives to a black hole. We saw that in the case of NGC 4258, it remained possible to fit a fermion ball (a neutrino star, for example) with the observed mass of 40 million solar masses inside the inner edge of the ring of gas (one-half light year) evidenced by the water vapor masers. The size of such an object could be as small as 0.02 light year and still be consistent with the upper limit (the critical mass) of ten billion solar masses required for the black holes in the most luminous quasars and radio galaxies. But given the mass–radius relationship for fermion balls ($R \propto 1/M^{1/3}$) the radius of a four million solar mass neutrino star would be even larger at 0.04 light year. In that case, the orbit of the star S0–2 reaching a minimum radius of 0.0017 light year would actually dip well into the neutrino star. The gravitational potential would be far from that of a point mass and the orbit would differ very significantly from Keplerian. The object at the center of the Galaxy is almost certainly a black hole.

10.5   Youth, Even More Paradoxical

In 2009, observations by the UCLA group confirmed that many of the young stars within 10 light years of the Center are in a clockwise rotating thin disk. They did not confirm the earlier claim by the Munich group that there is a second counter-clockwise disk, so the presence of this second disk remains controversial. But in any case, the planes of neither of the purported disks coincides with the other, nor with the Galactic plane, nor with the plane of the gas filaments comprising the northern arm of the mini-spiral. Moreover, the disks, when projected onto the plane of the Galaxy, rotate counter to the direction of Galactic rotation. This suggests that the disks are formed in some sort of random event, not in a smooth flow connected with Galactic rotation.

An infalling cloud provides a possible explanation (see Chapter 9). The cloud is stretched into a long filament that self-collides at high velocity, possibly leading to star formation in the resulting strong shock waves. The newly formed stars would lie in the original orbital plane of the infalling gas cloud, that is to say, in a disk. The fact that neither disk of young stars coincides with the plane of the northern arm (and rotate in the opposite sense) implies that these could not be the stars formed in any such infall event connected with the presently observed filament. This is not too surprising because the stars are at least one million years old and the northern arm, even with precession over several orbit times, is not more than 100 000 years old. In the context of this scenario these disk stars must have been formed in earlier such accretion events.

The stars closer to the center, the "S" stars for which the orbital parameters have been measured, do not appear to lie in any disk; their orbital planes are randomly distributed. The problem is that several of them pass very close to the center, within a few light hours for S0–2. Thus the tidal forces acting on any cloud that might collapse to form such a star would be extreme indeed. The constraint on the density of a cloud that could survive these extreme tides is comparable to the mean density of matter inside this innermost point in the orbit – more than 10^{15} atoms per cubic centimeter. Such high compression is asking a great deal of colliding streams of gas with initial densities of ten thousand particles per cubic centimeters, even at collision velocities in excess of 1500 km/s.

What could solve this problem of star formation so near the black hole? Perhaps there has been orbital migration after the stars were formed further out, possibly caused by instability in a disk. Perhaps these are old rejuvenated stars – old stars that have increased their mass through coalescing collisions in the very high stellar density environment near the center (although the high orbital velocities near the black hole would likely lead to disruption rather than coalescence). There is also the possibility of tidal compression of a cloud: the cloud passing near the center, as we see in Figure 9.6, is drawn out into a thin filament; the stretching occurs in the direction of motion, but in the other two directions (perpendicular to this motion) the cloud contracts.

The currently favored scenario involves rapid orbital evolution due to disruption of binary star systems in the tidal field of the black hole, an idea originally proposed by Jack Hills in 1988. In the local neighborhood (in the Galactic plane near the Sun), roughly half the stars are formed in binary pairs – two stars in orbit around each other. If this is also true for star formation in the central region of the Galaxy, then a certain number of close binaries will be disrupted by the tidal force of the black hole. Such an event would lead to one of the binary companions being more tightly bound to the black hole while the other is ejected with very high velocity. The tightly bound stars would comprise the S stars, and the ejected stars should be observable as hypervelocity stars in the outer halo of the Galaxy – stars having velocities in excess of 1000 km/s. The recent observation of several such hypervelocity stars by Warren Brown and collaborators lends credibility to this model. But although this mechanism eases the paradox of youth, it does not resolve it. It is still necessary to form young stars quite near the black hole.

10.6   Pele and the Alicanto

There are problems with all of these proposed mechanisms for star formation near the black hole, but the observations do suggest distinct and random star forming events very near the black hole – separate events perhaps connected with separate accretion events onto the black hole. It is certain that the black hole, at present, is not shining very brightly. The total power of Sgr A* over all wavelengths of electromagnetic radiation, from radio to X-rays, is less than 100 times that of the Sun. This is far less than the maximum possible power of a four million solar mass black hole which is given by the so-called Eddington limit (the same Eddington who was an early supporter of Einstein's theory of General Relativity, Chapter 3). The Eddington limit is due to the fact that when a compact object accretes too rapidly, the force from the radiation on the accreting matter overwhelms the gravity force which stops further accretion. The limit is proportional to the mass of the accreting object and for a four million solar mass black hole would be about 100 billion solar luminosities, comparable to that of the entire Milky Way Galaxy. So the black hole is far less luminous than it could be.

But presumably the Milky Way black hole grew from a smaller mass, so it has certainly accreted in the past. Perhaps it is like Pele of the Hawaiian islands in her struggle with Poliahu, the snow goddess. Mostly the mountain is quiet and cold but occasionally erupts and spews fire. Or perhaps it is more like the sleeping Alicanto of the Atacama desert. Occasionally it awakes and feeds and then shines like gold or silver and bright stars emerge from its eyes. Only in these periods of activity is it directly observable by the attentive miner.

If the black hole accretes at its maximum, the rate set by the Eddington limit, then it needs to be turned on only for 1% of the time to grow to its present mass in the age of the Galaxy; this would be its "duty cycle." For example, the black hole could be turned on for ten thousand years out of every one million years (Figure 10.9). Then it is possible that the newly formed stars are associated with these accretion events. In fact, it may be that the stars so near the black hole are formed *during* the accretion events. Perhaps the extreme radiation pressure on small dusty clouds could compress the clouds and thereby overcome the tidal forces and trigger star formation. Perhaps, during periods of activity, jets from the black hole could compress clouds sufficiently to trigger star formation.

With respect to the Galactic Center where the tides and luminosities are extreme, this is all very speculative, but it is possible that there is a connection between star formation and powerful outbursts of the massive black hole. This is

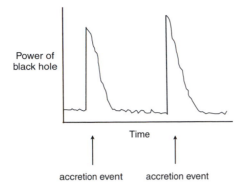

Figure 10.9 A schematic view of the power (luminosity) of the Galactic Center black hole over time. There are discrete outbursts corresponding to individual accretion events (infall of a cloud for example) but the black hole is very active for only 1% of the time. For example, the interval between outbursts could be one million years, but the length of an outburst is only ten thousand years.

why it is so important to understand the formation of young stars at the center of the Galaxy.

10.7   Summing Up: What Have We Learned?

When Horace Babcock published his famous paper on adaptive optics in 1954, it all seemed quite incredible that such a system could be designed and built – that the shape of the atmospherically disturbed wavefront could be detected and corrected many times in 1 second. This required detectors, computing facilities, and optical components that were barely imaginable at that time. The rapid increase in computing power (Moore's law) and the cold war involvement of the U.S. military in satellite detection changed all of that over the following forty years. By the time that the 10-meter class telescopes, the Keck and the VLT, came online the adaptive optics systems were ready to be applied, and use could be made of the full resolving power of these large telescopes.

The astronomer enters at this point and her or his skill is in knowing how to use these powerful new tools, what to use them for, and how to interpret the results (in some cases the astronomers actually build the instruments). It was that way with Oort when radio astronomy was developed and the 21-cm line was discovered; it was that way for Genzel and Ghez and their groups when AO became available on large telescopes. In retrospect it seems obvious that such new facilities should be applied to infrared observations of faint stars in crowded star fields near the Galactic Center; in practice it took enormous foresight, an ability to recognize and focus on the most important problem, and a forceful drive to get the best science out of the available technology.

The results have been astounding. There really is a four and a half million solar mass black hole in the center of the Galaxy; all reasonable alternatives have been eliminated. And even with all of the sophisticated modern technology, this result emerges from the most basic and traditional form of astronomy: positional astronomy, astrometry, of the sort practiced by Tycho and Kapteyn.

We have learned that there are one or two disks of young stars between 0.10 and 1 light year from the black hole; that these disks lie in apparently random angles to the Galactic plane, suggesting that they were formed in stochastic (random) accretion events such as the infall of small clouds from random directions. Within these disks, within a distance of 1 light month from the center of the Galaxy, there are the S stars, also young stars, with orbits that are being precisely traced. The compact radio source clearly coincides with the center of attraction, the center of mass, of the orbiting stars.

These star orbits in the central light months are, to the limit of current precision, perfect Keplerian ellipses. With accumulating data and improved optical systems, the orbit determinations will become more precise (the precision of astrometric observations always improves with time). At some point orbital precession due to the deviation from a point mass potential (a deviation from inverse square force) resulting from the presence of the background distribution of stars will be detected. And when that effect is removed, there will be precession of the ellipses due to General Relativity, the fact that the gravitational field is not perfectly Newtonian in the strong field limit – the same mechanism giving rise to the precession of the orbit of Mercury.

For a perfect four million solar mass Schwarzschild black hole (spherically symmetric) the predicted precession of orbit S0–2 would be about 14 arc minutes per revolution, that is to say 14 arc minutes per 16 years. For Mercury this is less than one arc minute per century, so the relativistic strong field effects near the black massive black hole should be much more obvious. Of course, the black hole in the Galactic Center has certainly swallowed angular momentum as well as mass, so it will actually be a Kerr black hole, but the Schwarzschild solution gives an idea of the magnitude of this relativistic effect. The effect, when observed, will constitute another test of General Relativity and will provide an estimate of the angular momentum of the black hole.

The question naturally arises: Who discovered the black hole in the Galactic Center? Bruce Balick and Robert Brown discovered the compact radio source, Sgr A* that has been identified with the black hole. This discovery was enormously important because it was suggestive of an unusual object and because it identified the point of interest in the central region. But there was no proof that it was a black hole – just a special radio source in a very special location. Don Backer and Dick Sramek, followed by Mark Reid and his collaborators, established that Sgr A* had no proper motion after subtracting that due to solar motion (Galactic rotation primarily): that the object appears to be at rest in the center of the Galaxy, but this is also no definite proof that it is a massive black hole and constitutes no measurement of the mass.

The discovery of young stars near Sgr A* (primarily by David Allen and his collaborators) led to doubts that there is a very massive object connected with Sgr A* because of difficulties forming stars in the strong tidal field of a million solar mass black hole and because all of the luminosity of the central region can be provided by these stars. But this is an indirect no-go argument. The final argument had to be dynamical – from the motion of gas or stars in the central region.

The gas filaments, the mini-spiral, do have motions that are consistent with a massive object, as shown by Serabyn, Lacy, and their collaborators – work inspired and supported by Charles Townes at Berkeley. But gas is subject to non-gravitational forces. What is needed are more perfect test particle tracers of the gravitational field. (As it turned out, the observed gas motions did give the about the correct result.) And such luminous test particles, discovered by David Allen, are provided by the young stars inside the inner light year for which all three components of the velocity can be measured – the velocity along the line of sight from the spectral lines, as well as the velocity perpendicular to the line of sight, the proper motion.

The orbits of individual stars can now be traced to an accuracy of 0.2 milli-arc seconds with adaptive optics on 10-meter telescopes. This has been done by groups headed by Reinhard Genzel and Andrea Ghez, who have demonstrated, beyond doubt, that such a bizarre object – an originally obscure consequence of the Schwarzschild solution to Einstein's field equation, hiding a mathematical singularity behind its horizon – does in fact exist in the Center of the Galaxy. And in other distant galaxies similar but even more massive black holes are the power plants behind the most luminous objects in the Universe, the radio galaxies and quasars.

11

Black Holes Here, Black Holes There...

11.1   Nearby, "Normal" Galaxies

In 1964 shortly after the discovery of quasars, George Field, then at Princeton, was impressed by the fact that the high redshift of quasars implied that they must be prevalent in the past – that very possibly they were associated with galaxy formation and not the end of a long process of galaxy evolution. He wrote a paper, now largely forgotten, entitled "Quasi-stellar radio sources as spherical galaxies in the process of formation." This was before the popularity, or even the respectability, of the black hole scenario, so Field's picture was that a collapsing protogalaxy could form a dense spheroidal cloud in which star formation would be rapid. This would lead to a quasar luminosity and flux variations due to frequent supernovae. He even pointed out that the density of quasi-stellar sources could be consistent with the present observed density of spheroidal galaxies. In many ways this was a very prescient contribution, but now we know that the energy source is likely to be accretion onto black holes, and the black holes do not disappear. They remain in the center of most reasonably massive galaxies as a generally quiet quasar remnant (Lynden-Bell's idea), waiting there to flare occasionally when a morsel of food drifts by.

But what is the evidence for this scenario? A black hole of four and a half million solar masses exists in the nucleus of the Milky Way; insofar as we understand gravity in this limit of very strong gravitational fields; we can now take this as an established fact. But how general is the presence of massive black holes in galactic nuclei? Is there evidence for dark mass concentrations in the centers of nearby galaxies that show no signs of present activity?

We have seen that a black hole of about 40 million solar masses is also present in the center of NGC 4258, a not-quite normal spiral galaxy. This is evident from

Figure 11.1 The great spiral galaxy in Andromeda, M31, a nearby companion to the Milky Way. The small galaxy above and to the left of M31 is the dwarf elliptical companion, M32. Both M31 and M32 apparently contain massive black holes in the center, but the larger galaxy, M31, has the larger black hole.

the water vapor masers in the dusty ring of gas orbiting at about one-half light year from the center of this galaxy – the source of the anomalous spiral arms seen in the radio continuum and now interpreted as jets near the plane of the galaxy. But what about other nearby galaxies without such anomalies?

The nearest large galaxy, at only about two million light years, is the great spiral in Andromeda, M31, a companion of the Milky Way (Figure 11.1). This galaxy has several satellites, smaller galaxies in orbit about it, including a dwarf elliptical galaxy, M32.

The nucleus of the Andromeda galaxy in visible light has been observed at high resolution since the early 1970s when it was first imaged from a balloon in the upper atmosphere – the Stratoscope, a project of Martin Schwarzschild and his colleagues at Princeton University. At resolutions of a few tenths of an arc second, it appeared to be asymmetric. Then, in the early 1990s the Hubble Space Telescope revealed what seemed to be two nuclei separated by one-half arc second or about 5 or 6 light years in projection; this double structure is shown in Figure 11.2.

There is very little interstellar gas in the central region, so the only tracer of the mass distribution is the motion of the stars. In 1988, Alan Dressler (Mt. Wilson) and Douglas Richstone (Michigan) and, independently, John Kormendy (Victoria) measured the kinematics of the stars in the nucleus from absorption lines arising from elements such as calcium in the atmospheres of the old stars; their results are shown in Figure 11.3, which is the measured rotation of the system of stars

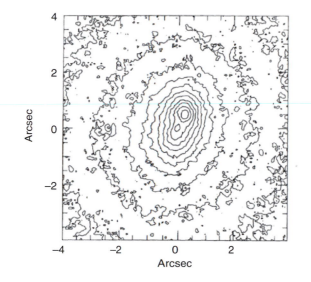

Figure 11.2 Distribution of visible light in the center of M31 in the inner few arc seconds. The double nucleus is obvious. (From Lauer et al. 1993.)

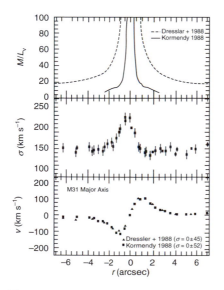

Figure 11.3 Kinematics of M31. The lower panel is the rotation velocity as a function of distance in arc seconds from the center of this galaxy; the middle panel is the velocity dispersion; and the top panel shows the mass-to-light ratio implied by solving the Jeans equation. At the distance of M31 (2.3 million light years) 1 arc second corresponds to 11 light years. The points are from Dressler and Richstone (1988) and Kormendy (1988). (The figure is given by Kormendy and Richstone 1995).

and the spread in their random velocity (the velocity dispersion) as a function of distance from the center. Applying the Jeans equation to these observed stellar kinematics implies that a central mass of at least thirty million solar masses is present. The calculated mass-to-light ratio interior to a given radius is shown also in Figure 11.2, and we can see that there is clear evidence for a dark mass at the center.

But there are apparently two centers in the nucleus of M31, so which is the real center? The velocity dispersion appears to peak on the fainter of the two nuclei in the center, and we take that to be the true location of the massive object. Then what is the other apparent nucleus in M31? A plausible proposal by Scott Tremaine (1995, then at Toronto) was that this arises from a disk of stars with an elliptical shape. If the gravitational force is that of a point mass (Keplerian) then, like the orbits of planets in the Solar System, the orbits of the stars will be ellipses with their long axes fixed in space. If the long axes are all in the same direction, then the disk will appear elliptical, and the second nucleus, or bright spot, is due to a concentration of stars at the point most distant from the center of the mass distribution; the stars are more concentrated there because they move more slowly farther from the point mass (Kepler's second law). But this places a constraint on the mass of the central object: it must be more massive than about twenty million solar masses if the gravitational field, including that of the surrounding stars, is to be sufficiently Keplerian – that of a point mass. Otherwise, the elliptical shape would vanish on the time scale of several orbital periods (about 30 000 years) because the orbits (the long axes) precess; those of stars near the center would precess faster than those further out, so the shape of the disk would rapidly become rounder.

The scale of this proposed disk would be on the order of the separation of the two nuclei – 5 to 10 light years. This is reminiscent of the CND (the circumnuclear disk) in the center of the Milky Way. Perhaps it has the same origin. For example, tidal distortion of an infalling gas cloud and subsequent star formation, except that it must have been formed long ago because the stars are old; they have the red color of the majority of the bulge stars.

The similarities with the Milky Way nucleus do not end there. In the near vicinity of the mass concentration, within 1 light year, there appears to be a cluster of blue stars, stars younger than 100 million years. They are certainly near the mass concentration because their velocity dispersion is quite large – several hundred kilometers per second. So star formation apparently occurs in bursts near the black hole of M31, as in our Galaxy. Star formation near the black hole in M31 has also found a way of getting around the problem of tides. Perhaps such star formation is a general phenomenon connected with outbursts of the black hole.

The latest estimates for the mass of the black hole in M31 range from 75 to 100 million solar masses, that is to say, about twenty-five times larger than that of the Milky Way black hole. But given the ambiguity in the isotropy of the velocity field (is the spread in velocity the same in all directions?) how certain can we really be that this is actually a black hole from the stellar kinematics alone? Lacking individual orbits of stars that pass within a fraction of light year of the center, the extreme concentration mass cannot be as certain as in the nucleus of the Milky Way. If, for example, there were a population of stars on extreme radial orbits that pass very near the center, then this would give the appearance of an increasing spread in velocity along the line of sight without a mass concentration. But modeling of the run of velocity dispersion by a number of independent researchers using a variety of methods and assumptions finds no evidence that we are being fooled. There is most likely a black hole of mass near 100 million times that of the sun residing in the nucleus of M31.

Also M32, the dwarf elliptical satellite of M31, shows the same kinematic evidence for the presence of a mass concentration even though it is twenty or thirty times less luminous than it large companion. In 1984 John Tonry, then of MIT, found that the velocity dispersion in this gas-free galaxy also increased dramatically in the central region; that is, there is a large implied central mass-to-light ratio. Later higher resolution observations and more thorough analysis implied a mass concentration of 2.5 million solar masses, similar to that in the Milky Way but a factor of thirty times smaller than that in its parent, M31.

In the 1990s observations of stellar kinematics in other more distant galaxies as well as nearby objects began to demonstrate the same pattern – there were apparent mass concentrations without light concentrations – dark mass – at the very centers. These were almost entirely elliptical galaxies or spirals with conspicuous bulges. Black holes seem to have an affinity for galaxies with substantial spheroidal components, not for pure disk systems. At present, there are more than fifty galaxies, mostly within distance of 100 million light years, with the kinematic signatures of black holes and estimated masses up to ten billion solar masses. This is a sufficiently large sample that systematic trends should become evident.

11.2   Bigger Black Holes Live in Bigger Spheroids

In the year 2000 two papers appeared in the same issue of the *Astrophysical Journal Letters*, both announcing a very striking result: there appeared to be a fairly tight correlation between the mass of the central black hole in spheroidal stellar systems, elliptical galaxies, or spiral galaxy bulges, and the observed spread in velocity of the stars comprising the spheroid – the stars well away from the region where the black hole dominates the gravitational field. In other words, the more massive the black hole, the larger the velocity dispersion of the stars in the spheroid. The first paper, by Laura Ferrarese (UCLA) and David Merritt (Rutgers), claimed that for twenty-six galaxies covering a range of 1000 in black hole mass, that mass was proportional to the observed line of sight velocity dispersion to the power of 3.8 (i.e., $M_{BH} \propto \sigma_{los}{}^{3.8}$) with a scatter about this relationship of a factor of 2.5. The second paper (which followed immediately in the *Astrophysical Journal*), by Karl Gebhardt (UCSC) and a large group of co-authors (known as the "nukers"), made a similar claim on the basis of about twelve galaxies (the power law was a bit larger with an exponent of 4.8).

In some sense this was not surprising because, previously in 1998, John Magorrian (Toronto) and colleagues (the nukers) had found that that the black hole mass appeared to be proportional to the luminosity of the galaxy spheroid – the more stars, the more massive the central black hole ($M_{bh} \propto L_{sph}$). Moreover, it had been known for some time that there was a correlation between total luminosity and velocity dispersion of elliptical galaxies, the Faber–Jackson relation in which the stellar luminosity of the galaxy is roughly proportional to the fourth power of the stellar velocity dispersion ($L_{sph} \propto \sigma^4$). So both correlations together would imply that the black hole mass should also increase as a high power of the velocity dispersion. But both of these new papers claimed that the correlation of black hole mass with velocity dispersion was actually tighter (had less scatter) than that with luminosity – that this new correlation appeared to be more fundamental. In either case, this was strong evidence that the black hole mass was related to the properties of the host spheroid – bulge, or elliptical galaxy. The updated form of this correlation is shown in Figure 11.4.

How does this correlation come about and what does it say about black hole formation? A simple answer is that the surrounding spheroidal system is the source of gas for the growth of the black hole. Stars, even the old stars found in elliptical galaxies, lose mass during their lifetimes. This mass loss is arises from a variety of mechanisms: stellar winds, planetary nebulae (a nonexplosive expulsion of envelopes of red giant stars), and type I supernovae (exploding stars in binary systems). The overall rate of mass loss is uncertain, but a reasonable estimate is 0.1 to 0.2 solar mass per year per 10 billion solar luminosities. This means that a spheroidal system of 10 billion solar luminosities would lose about 1.5 billion solar masses during the 10 billion year lifetime of the galaxy. This is quite a large quantity of gas.

What happens to all of this gas that is shed by the old stellar population? It is certain that it does not all go into the formation of new stars, at least not with a normal distribution by mass; that would be obvious in the colors and spectra of the spheroid (it should contain a conspicuous blue component). The standard

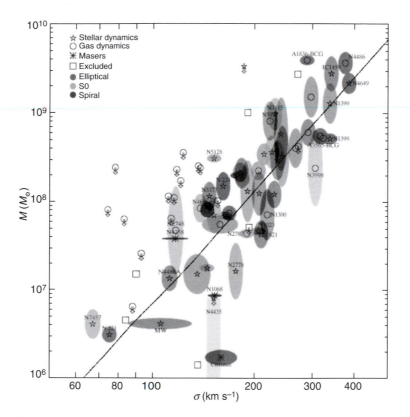

Figure 11.4 The relationship between the mass of the central black hole and the line-of-sight velocity dispersion in a spheroidal galaxy (elliptical or bulge). This is a log–log plot so the straight line relationship is actually a power law with an exponent near four. (From Gültekin et al. 2009.)

dogma is that the gas is further heated by supernovae and leaves the galaxy in the form of a steady hot wind. But this depends on the efficiency of the supernovae in heating the gas. It is also possible that, at least in the central regions, there is a cooling flow; the gas cools and flows inward at a steady rate. It is also possible that both are occurring simultaneously with gas in the inner region flowing in and in the outer region flowing out. One thing is certain – in larger galaxies, with higher luminosity and therefore more stars, more gas is created, and a higher mass of gas is likely to flow inward. From the numbers above we see that only 5% of the gas evolved from hot stars need flow inward and contribute to build up of the black hole over 10 billion years in order to account for the relationship of black hole mass with spheroid luminosity.

This is not, however, the standard view of the correlation. The now conventional picture involves a phenomenon known as "feedback": the black hole itself

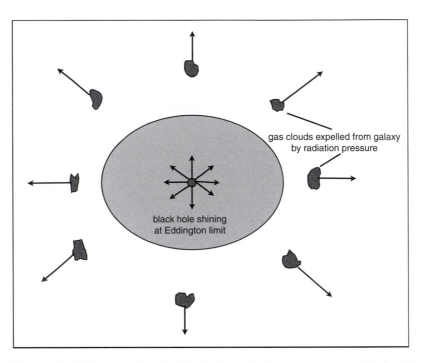

Figure 11.5 Schematic sketch of feedback mechanism. A massive black hole at the center of a spheroidal galaxy shines at its Eddington limit with sufficient luminosity to blow gas out of the galaxy. Thereby star formation ceases in the galaxy, and this limits its mass in stars.

limits its own growth and that of the surrounding galaxy due to its outbursts, which blow out gas that might accrete or form stars. One might naturally ask how the black hole which is such a relatively small component of the galaxy, less than 0.2% in mass, can have such a dramatic effect on the enormous galaxy around it. That is because the radiant *energy* emitted by the black hole during its growth is several thousand times larger than the total gravitational energy of the entire galaxy; if only 0.1% of the energy emitted by the black hole is absorbed by the gaseous component of the galaxy, the black hole can have a drastic effect on the resulting evolution of the galaxy by blowing out the gas from which the stars of the galaxy forms and the black hole grows. Feedback could tune the mass of the galaxy to the mass of the black hole.

Joe Silk and Martin Rees (Cambridge) proposed in 1998 that if some small percentage of the energy produced by the black hole is absorbed by the gas in a galaxy (they suggested that the interaction occurs through a hot wind from the central quasar), then there would be a relation between the black hole mass and the spheroid velocity dispersion (this was before the relation was discovered!).

Requiring that the black hole have sufficient mass (or Eddington luminosity) to expel gas from the galaxy (the expulsion velocity should exceed the escape velocity) means that the black hole mass should be proportional to the fifth power of the galaxy velocity dispersion ($M_{bh} \propto \sigma^5$), not far from the observed relation. In 2002, Andy Fabian and collaborators (Cambridge) suggested that gas could be driven out by the luminosity of the quasar itself, by means of radiation pressure directly transferring momentum to a cloudy, X-ray–absorbing interstellar medium farther out in the galaxy. This process would lead to the black hole mass being proportional to the fourth power of the stellar velocity dispersion. In either case, the principal relation would be between the black hole mass and the stellar velocity dispersion without the intermediary of the spheroid luminosity or mass in stars.

There is another important motivation for considering the mechanism of feedback. In the current favored Lamda-CDM cosmology (a universe dominated by a cosmological constant and cold dark matter), galaxy formation is hierarchical: big galaxies form from the merger of smaller galaxies. In the context of this picture of galaxy formation (or "assembly" as it is presently called), the distribution of galaxy dark halos by mass can be calculated through numerical simulations. It turns out that the simulations produce too many massive halos, more than observed if all the normal matter, the baryons, in these halos go into forming stars. So a fix for this perceived problem is provided by active galaxy feedback. When it becomes active, the black hole at the center of a massive galaxy blows away the remaining baryons and shuts off the growth of the visible galaxy. This mechanism can work for reasonable choices of the several free parameters of feedback, such as the efficiency of coupling the luminosity of the black hole to the gas in the galaxy (see Croton et al. 2006 for a discussion of this process).

Although feedback is the favored mechanism for explaining the correlation, there are other ideas. In the scenario of hierarchical galaxy formation, each of the merging small galaxies may contain a central black hole with a mass that can vary over a large range; Then after merging, the black holes will find their way to the center of a single larger galaxy (via dynamical friction) and will also merge, producing a larger black hole. If this happens a number times, this process will actually lead to a fairly tight relation between the final black hole and galaxy spheroid mass (proposed by Knud Jahnke and Andrea Macciò in 2011). This is a statistical rather than causal explanation of the correlation.

It is important to identify the correct mechanism because this can distinguish between the various scenarios for galaxy-black hole formation and coevolution. So to find evidence for (or against) feedback we need to look back at most active galaxies – the quasars and the epoch near redshift of two when the quasars are shining brightly and the black holes are growing rapidly.

11.3   Blast from the Past

One thing is certain – the black hole in the center of the Milky Way was never a quasar because it is not massive enough. At a few million solar masses its Eddington luminosity is *only* about one hundred billion times that of the Sun, about equal to the total power of the Milky Way, whereas quasar luminosities can range up to 10 000 Milky Way luminosities. The Milky Way certainly could have been (and may be again) a Seyfert galaxy, but this does not put it in the range of typical quasars. But then its bulge mass and therefore its black hole is rather small, near the bottom of the plot of black hole mass versus velocity dispersion (Figure 11.3). Many of the galaxies in the upper half of this figure could well have had quasar luminosities. Is the density of such galaxies comparable to the density of old quasars – of dead quasars? This was, after all, Lynden-Bell's original proposal.

Given a plot like Figure 5.3 that shows the number density of quasars as a function of redshift (or cosmic time) we can estimate the present density of dead quasars. Taking the typical quasar black hole mass to be 100 million solar masses, that turns out to be near 0.0001 (one ten-thousandths) per cubic million light years (this is equivalent to an average separation of dead quasars of roughly 15 million light years). Then given the relationship between black hole mass and bulge velocity dispersion or luminosity and the distribution of galaxies by luminosity we can estimate the number of galaxies that could have hosted a quasar (with black hole mass greater than 100 million solar masses). That also turns out to be about 0.0001 per cubic million light years. This coincidence of numbers is a remarkable confirmation of Lynden-Bell's original idea: old quasars really do appear to reside in the nuclei of massive galaxies, and they are to be identified with massive black holes. They are presently not shining so brightly because they are on a starvation diet – they are underfed (this estimate was first made by Andrzej Soltan in 1982).

But they were certainly not underfed at a redshift of two when the Universe was about 25% of its present age (or about 3.4 billion years old) and massive black holes were shining brightly. This period of quasar activity also coincides with the peak rate of star formation in the Universe. The star formation rate, estimated from colors of high redshift galaxies, appears to reach a maximum between redshifts of two to three, at about the same cosmic epoch when massive black holes in galactic nuclei were experiencing their pubescent burst of growth. In other words, black holes and the stellar content of galaxies build up at the same time; they coevolve.

This in itself does not imply a causative relationship. Specifically, it does not necessarily mean that black holes quench star formation by blowing the gas away (the star formation itself could also be instrumental in removing the gas), and it may be that these two components of massive galaxies are simply forming at the same time. But there does also appear to be some more direct evidence of feedback due to active nuclei. Clusters of galaxies very often contain a central active galaxy. Such clusters also contain hot gas, but in those with active galaxies it is generally the case that the cooling timescale for the gas (the time it takes for the gas to cool by radiation in the form of X-rays) is shorter than the age of the Universe. So the hot gas should be cooling and flowing into the center of the dominant galaxy – the active galaxy. The problem is that this cool gas is not observed; the temperature near the center of the cluster does not drop conspicuously. So what happens to the cool gas that should be there?

The current popular idea is that this gas is reheated by the active galaxy at the center, and there indeed is some evidence, at least in several cases, that this is happening. The Perseus cluster, for example, has an active galaxy, NGC 1275, at its center. There are clear signs of interaction of the radio source associated with this object (Perseus A) with the hot gas in the cluster. Figure 11.6 shows the distribution of hot X-ray–emitting gas (from the work of Andy Fabian and colleagues); there are holes in the gas distribution that coincide with the radio emission produced by the active galaxy, presumably the black hole. There are also shock waves and ripples (sound waves) that emanate one-million light years

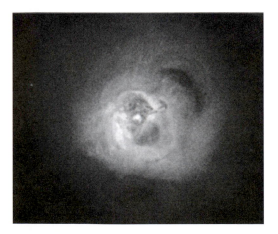

Figure 11.6 The X-ray emission from the Perseus cluster of galaxy that traces the distribution of hot gas with the darker regions indicating less gas. The image spans 300 000 light years across the central region of the cluster. The two holes at the near the center coincide with the radio jets from the active galaxy NGC 1275. Farther out there are ripples in the hot gas distribution apparently caused by the radio jet from NGC 1275. This is a clear indication of mechanical input of energy by the jets into to the surrounding medium. (Image from NASA/CXC/IoA/A.Fabian et al.)

out from the galaxy. The galaxy, or its black hole, is clearly having an impact on the surrounding medium far from the object itself – a direct mechanical input of energy from the radio jet – the sort of impact that could stop the cooling flow and block the further build up of the stellar content of the galaxy as well as the growth of the black hole.

Of course, given the vast reservoirs of hot (but cooling) gas, clusters of galaxies are rather extreme environments. But there are other more direct signs that such feedback may occur on the scale of individual galaxies – that energy may be directly transferred from a radio jet to gas motions. For example, in 3C 293, a relatively nearby radio galaxy, Bjorn Emonts and colleagues at the Kapteyn Institute and Dwingeloo, found an absorption line in neutral hydrogen (the 21-cm line) – absorption against a region of enhanced radio emission from a distorted radio jet. The absorption has a long wing extending out to redshifts 1000 km/s less than that of the galaxy itself (see Figure 11.7). Because an absorption line clearly arises between us and the radio source, this means that the gas is flowing out of the galaxy at high velocity, possibly being driven by this jet interaction.

There is evidence of such gas outflow in several quasars, including very old quasars at redshifts well in excess of three. For example, R. Maiolino (Cambridge) and colleagues recently found a quasar at a redshift of 6.4 (meaning that the object was in place when the Universe was only 895 million years old) with a very wide spectral emission line of ionized carbon emission. The width of the line implies

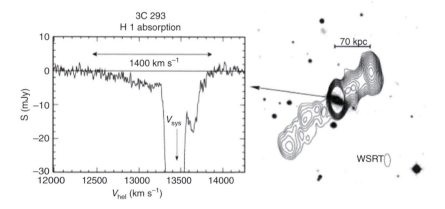

Figure 11.7 The left-hand panel shows the 21-cm absorption line against the radio continuum source about 30 000 light years from the center of this powerful radio galaxy. The radio continuum on a larger scale is shown in the right-hand panel. The wing of the absorption line extends more than 1000 km/s to blue (i.e., to lower redshifts) of the systemic recession velocity of the galaxy. This means that the gas is moving out of the galaxy with high velocity, possibly due to an interaction with the radio emitting jet. (From Emonts et al. 2005.)

that there is gas moving out of the galaxy again at a velocity higher than 1000 km/s, and that it is a substantial amount of gas: more than 3500 solar masses per year are being lost from the host galaxy. This is consistent with the feedback scenario, but what is surprising is that the massive galaxy host as well as the massive central black hole are already in place at such an early epoch. So the question arises: How do massive black holes form in galaxies so soon after the Big Bang?

11.4   The Early Formation of Black Holes

We have seen that there are a number of paths to gravitational collapse – there are many ways of forming a massive black hole. But if the hole forms by accretion of gaseous disk, then it can grow only at a rate allowed by the Eddington luminosity or about one galaxy luminosity (100 billion solar luminosities) for a black hole of one million solar masses. For a 1 million solar mass black hole this amounts to a growth rate of about 0.02 solar mass per year with a 10% efficiency of converting rest mass into energy. Starting with an initial black hole mass of ten solar masses (a typical black hole mass that might emerge from normal processes of stellar evolution and collapse) and accreting continuously at its Eddington rate, the black hole would grow exponentially; the mass doubles every 30 million years. This means that it would take about one billion years to grow to a mass of five billion solar masses typical of QSO black holes. A timescale of one billion years corresponds to the age of the Universe at a redshift of about six, the point at which luminous quasars with substantial gas outflow are observed.

Thus it might seem possible for a stellar mass black hole to grow to a super-massive black hole capable of producing quasar luminosities at redshifts greater than 5, but this cuts it rather close. The process can begin only after the first generation of star formation and requires that black holes are growing at their maximum possible rate. So this becomes a problem: How can massive black holes in galactic nuclei form and grow to their required mass so quickly on a cosmic timescale – when the age of the Universe is only one billion years? This timescale requirement would seem to challenge any scenario in which the black hole forms in a dense nucleus of stars, such as that outlined in Chapter 5, where single massive objects form by coalescing collisions between stars. The nucleus must be in place before the growth of the black hole begins, further straining the timescale problem. But it remains a mechanism that should be considered; perhaps a very dense nucleus of stars can form in the initial collapse of the protogalaxy.

One process by which black holes can grow in dense galactic nuclei was suggested by Jack Hills in 1975. Recall from Chapter 3, freely falling Sally can pass safely through the Schwarzschild radius if the black hole is more massive than about one million solar masses. If the hole is less massive then she is torn apart by tides before she crosses this horizon. It is the same for stars, but here the mass limit is about 400 million solar masses. If the hole is smaller than this limit, then it is too small to eat stars in a single gulp; the star that wanders too close to the black hole will be torn apart and its matter will be consumed in gaseous form – presumably in an accretion disk. This is good because it produces the radiation of an accretion disk; the black hole can shine and produce an active galactic nucleus. But for black holes more massive than this 400 million solar masses, the star is swallowed whole and not disrupted. No electromagnetic radiation is produced. This is also good because we know that after a rapid period of shining brightly, black holes turn off at some point.

A particular realization of this scenario is shown in Figure 11.8. Here, starting with a 1000 solar mass seed black hole in a particularly dense nucleus of about three million solar masses per cubic light year, we see that the hole mass and luminosity grow via tidal breakup of stars and accretion of stellar matter for

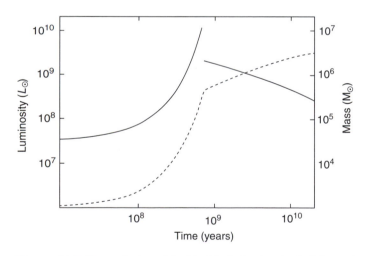

Figure 11.8 The evolution of the black hole luminosity (solid curve) and mass (dashed curve) for a model dense galactic nucleus of about three million stars per cubic light year and an initial seed black hole of 1000 solar masses. The black hole is growing and shining by tidally disrupting and consuming stars in the surrounding stellar system. This is a logarithmic plot with the time at the bottom given in years and the luminosity and mass given in solar units. The systems evolves to a peak luminosity after about one billion years with the mass consumption being limited by the Eddington rate. The excess mass is assumed to accumulate, but after one billion years this reservoir of gas from disrupted stars is depleted and the mass consumption and luminosity is due to the rate at which the black hole tidally disrupts stars. (From Sanders and van Oosterom 1984.)

the first billion years; near the peak luminosity the mass from disrupted stars is increasing faster than it can be consumed at the Eddington limit, so a reservoir of gas builds up, presumably in an accretion disk. But after one billion years that reservoir is consumed and the luminosity (and rate of growth of the black hole mass) decreases. Unfortunately, it does not appear that the model can achieve quasar levels at 100 000 times a galaxy luminosity) even with these rather extreme conditions

For this model to work, the nucleus would have to form rapidly in an already dense state, or dynamically evolve to that state via stellar dynamical processes involving two-body gravitational encounters, and these are generally slow. But in any case, the process of tidal disruption of stars by black holes in galactic nuclei with relatively small black holes, like that in the Milky Way or M32, must occur every few thousand years or so. The accretion disk formed by such an event would be rather short lived (probably less than 100 years), leading to a sudden flare-up in an otherwise normal nucleus. That can have observational consequences: continual surveying of a number of galactic nuclei (more than 1000 objects) would have a reasonable chance of catching such a flare. In fact, there is recent evidence reported by S. Gezari and colleagues (Baltimore) that such an event has been observed in the center of a distant and otherwise inactive galactic nucleus.

In general the early appearance of black holes with sufficient mass to power quasars – mass greater than one-billion solar masses – suggests that the formation and evolution of massive central black holes goes along with that of their host galaxies, or at least their spheroidal components.

The idea that engines of active galactic nuclei emerge during the formation of galaxies began with the suggestion of George Field, but now there is the additional requirement that a substantial concentration of mass must accumulate very early within its Schwarzshild radius. It could be that the initial collapse of matter that leads to a galaxy also leads to a massive seed – a black hole with an initial mass sufficiently large (greater than 1000 solar masses) that it can grow rapidly by accretion (in less than one billion years) to that required by quasar luminosities (greater than 100 million solar masses).

As discussed in a recent review by Marta Volonteri and Jillian Bellovary (Paris, University of Michigan), there are several barriers that must be overcome for this to happen: the collapsing gas will certainly have some angular momentum, and even a small residual angular momentum will be sufficient to stop the collapse at radii many times larger than the Schwarzschild radius; that is to say, collapsing gas will form a disk well outside its Schwarzschild radius. A self-gravitating disk supported largely by rotation is unstable – it tends to form a cigar shape or a bar and such structures can transport angular momentum outward, allowing gas to flow inward.

This inward transport of gas on a rapid time scale (an orbit time or less than 100 thousand years) could lead to a massive seed, but there is the problem of fragmentation. The gas, as it collapses and becomes denser, tends to cool, and the cooling gas tends to fragment into objects that would form stars, perhaps massive stars. Then the black hole seeds would be on the order of stellar masses, and this would probably promote the formation of clusters of black holes rather than a single massive black hole. The initial formation of stellar mass objects means that the eventual formation of a massive black hole must rely on the slow stellar dynamical processes (there are arguments that these processes might not be so slow after all, given the possibility of instability and rapid collapse of a dense cluster).

It is possible that the cooling of the collapsing gas (and subsequent fragmentation) is inhibited by the lack of elements heavier than hydrogen and helium. It is these heavier elements such as carbon and oxygen that are the principal agents of radiative cooling of warm gas in the Universe, but they are not present in the primordial gas from the Big Bang; these elements must be produced in the initial generation of stars (so-called population III stars). In one scenario, proposed by Mitch Begelman (Colorado), Marta Volonteri, and Martin Rees (Cambridge), the pristine collapsing gas cloud, missing these cooling elements, does not fragment but forms a "quasi-star," an object of one million solar masses with a black hole at its center surrounded by an extended gaseous atmosphere – sort of a super-super red giant star. As the black hole grows, the atmosphere expands and eventually disperses, leaving the bare black hole of thousands of solar masses behind. This massive seed could then rapidly grow to quasar proportions.

Basically, it is presently unknown how massive black holes emerge and grow in forming galaxies. There are a number of ideas, but these are difficult to test observationally given the need to observe young and very possibly small galaxies at high redshift. So none of these various ideas has yet been frozen into dogma, which is actually an exciting state of affairs for the imaginative theorist. However, there is a much closer laboratory where these theoretical constructs can be tested: ideas on episodic fueling and activity and its duty cycle; on the tidal disruption and accretion of stars or molecular clouds; on the impact of activity on the surrounding galaxy – that is, feedback, radiative and hydrodynamic; on star formation in extreme environments and the relationship between activity of the black hole and such star formation. This is a laboratory where even our fundamental theory of gravity in the limit of very strong fields can and will be tested. This is the laboratory of the Galactic Center, where our wildest theoretical speculations must confront, head on, the reality of observations.

12

Traces of Activity: Past, Present, and Future

12.1 Effects of Black Hole Outbursts on the Galactic Center Gas

Some catastrophic events on Earth have a long-term drastic impact on the geology of the planet and biological evolution. Others have dramatic but short-lived effects. Examples of the first category would be volcanic eruptions that can create mountains or islands à la Pele, and collisions with astroids or comets that produce enormous craters and destroy many life forms; such episodes create relics that can persist longer than the time interval between these events. Examples of events with short-term effects would be earthquakes and tsunamis that can be devastating locally but with few long-lasting consequences on the structure of the Earth (apart from accumulative effects of many such events). While a tsunami creates a ripple on the surface of the ocean that can flood and destroy vast distant coastal regions, the long-term global effects of a single such event are negligible.

Into which category would the hypothetical occasional eruptions at the Galactic Center fall? What is the effect of outbursts of the black hole on the surrounding environment? How long do such effects last and how far do they extend into the Milky Way? It is certain that the black hole is presently inactive – in fact, almost embarrassingly so. In 2001, using the Chandra satellite, F.K. Baganoff and collaborators detected X-ray emission from Sgr A*, quite possibly due to thermal emission from hot gas in the near vicinity of the black hole. The total power is about one solar luminosity, but (as was soon discovered) with occasional flares occurring with a frequency of about one flare per day having a duration typically of a few hours; some of these flares are 100 times more luminous than the average power. Infrared emission from Sgr A* was finally detected in 2002 by both the Max Planck and the UCLA groups. This was possible only with adaptive optics, which removed confusion with the nearby orbiting young stars. Here again there is short-term flaring activity, but the average quiescent power is about 100 solar luminosities (10^{35} ergs/s). So the present power of the black hole over all wavelengths is only that of 100 suns with most emerging in the near infrared.

The maximum possible power of a four million solar mass black hole, the Eddington luminosity, is about 100 billion times that of the Sun, equivalent to the entire power of the Milky Way emitted by normal stars. So the black hole is presently emitting a very small fraction of what it could do – only one-billionth of its maximum possible. With an efficiency of 10% for converting accreted mass into energy, this would correspond to an accretion rate of one-billionth of a solar mass per year (10^{-9} solar mass/year), which amounts to ten solar mass over the entire lifetime of the Galaxy. This low accretion rate is quite problematic because even the normal winds blowing from the massive young stars in the central region should provide a much larger supply of gas to the black hole (Wardle and Yusef-Zadeh 1992). In any case, the accretion rate must occasionally be much larger to construct a four-million solar mass black hole over the history of the Universe. Or the efficiency of mass to energy conversion must be one one hundred millionth, but then this would be rather strange because the motivation for black holes in galactic nuclei is to explain extremely high luminosities from very compact regions.

Most likely, the black hole at the Galactic Center exhibits occasional outbursts with a power going up to its Eddington luminosity – the power of a typical Seyfert galaxy. The accretion rate during such outbursts could be as high as 0.1 solar mass per year, implying that the entire mass of the black hole could be supplied in forty million years – less than 1% of the age of the Universe. So the black hole need shine at its maximum only for 1% of the time in order to accrete its observed mass – for example, ten thousand years out of every one million years. There could, of course, be lower luminosity outbursts which are longer in duration and more frequent, as implied by the X-ray flares.

What would such outbursts do to the surrounding environment? Is it possible to observe traces of past activity in the greater central region of the Galaxy – in anomalous motions of gas, for example. After all, the effect of very massive black holes in luminous active galaxies, the quasars, has been invoked to explain the high-mass cutoff in the distribution of galaxies by stellar mass. This is supposedly accomplished by driving gas away from the galaxy and cutting off star formation (there is observational evidence that this is happening in the more luminous quasars). The same process supposedly establishes the black hole mass–bulge velocity dispersion relationship in galaxies ranging from the Milky Way up to the massive elliptical galaxy hosts of very large black holes. If so, then we might expect that some traces of past activity of the black hole might be observationally evident in the gas distribution and motions in the central region of the Milky Way.

Let's take the most favorable case, energetically, for a hydrodynamical effect of the black hole on gas motions. Suppose that in one outburst the black hole emits at its Eddington luminosity, 100 billion solar luminosities, for ten thousand years once every million years. Then the total energy emitted over this outburst is about one hundred times the rest mass energy of the sun (by Mc^2) or 2×10^{56} ergs. Now most of the gas in the central 600 light years is in the form of molecules in clouds – the so-called central molecular zone. The total mass of these clouds is approximately 100 million solar masses and they are rotating about the center with a velocity of around 130 km/s, so the gravitational energy of these clouds is 2×10^{55} ergs or ten times the rest mass energy of the sun. So the total luminous energy emitted by the black hole in such a hypothetical outburst is ten times larger than the gravitational energy of the gas in the central molecular zone. If 10% of the energy output of the black hole could go into gas motions we might expect that such an outburst could have a dramatic effect on the kinematics of molecular clouds in the central region.

The problem is this 10%. How is the power output of the black hole coupled to gas motions with a 10% efficiency? One obvious mechanism is the direct pressure of radiation from the black hole on a dusty cloudy medium as is illustrated in the left-hand panel of Figure 12.1. The photons streaming away from the black hole carry momentum and the total momentum available for pushing the gas is just the energy emitted by the black hole over an outburst divided by the speed of light ($\approx E/c$). But this is only 1% of the outward momentum of the "expanding molecular ring" at a radius of 600 light years. So it appears impossible that the radiation emitted by the black hole could be the cause of this apparent expanding ring; there is just not enough momentum available in the radiation.

It could be that outbursts of the black hole have a more indirect effect on gas motions of the molecular gas. For example, if the black hole triggers star formation then this could lead to a temporarily enhanced rate of supernovae (illustrated in the right-hand panel of Figure 12.1). The exploding stars eject their outer layers and create an expanding shells of gas; in this way supernovae could have a mechanical input of kinetic energy into the molecular gas. Ten to one hundred thousand supernovae in an outburst of the black hole every one million years could supply a kinetic energy of up to 100 solar rest masses (10^{56} ergs), exceeding the total gravitational energy of the clouds in the central molecular zone.

Energetically then, it is possible that supernovae from star formation episodes could explain features such as the expanding molecular ring. However, most likely is the explanation already given – noncircular motions in the gravitational potential of the central bar that we know is present from the near-infrared maps of the central bulge of the Galaxy. There remains the possibility that gas motions

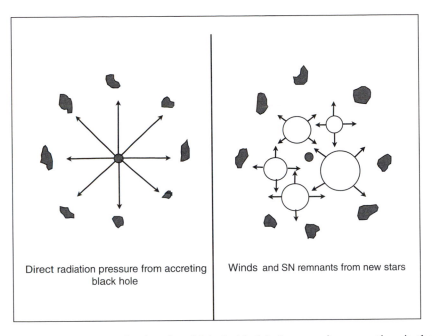

Direct radiation pressure from accreting black hole

Winds and SN remnants from new stars

Figure 12.1 Two mechanisms by which the black hole can excite gas motions in the surrounding medium. On the left-hand side the black hole exerts direct radiation pressure on the dusty clouds and accelerates them outward. The problem is that the total momentum available in an Eddington outburst seems insufficient to push the clouds up to velocities in excess of 100 km/s. On the right side, an outburst of the black hole triggers star formation, which then, via winds and supernovae explosions, exerts mechanical pressure on the molecular clouds. Energetically this is a possible mechanism.

closer in to the center may be affected by occasional outbursts of the black hole and/or resulting star formation. The massive molecular clouds in the inner two or three hundred light years of Sgr A* have large random velocities and a highly asymmetric distribution (most massive molecular clouds are found on one side of the Galactic Center). This very transient distribution may result from black hole activity and related star formation.

The circumnuclear disk (CND) is essentially a torus of dusty, turbulent molecular gas with an inner cavity of roughly 3 light years. It may be the principal structure that occasionally feeds the black hole. Mark Morris, Andrea Ghez, and Eric Becklin (UCLA) have argued that this may occur as a "limit cycle"; that the ring is held up in the galactic gravitational potential partly by winds and supernovae from newly formed stars and when these subside as the stars age, the CND will migrate inward and initiate a new black hole outburst and an episode of star formation that cuts itself off by blowing the CND outward again.

Another possibility is that the bar itself is instrumental in creating such cyclic behavior. Anthony Stark (Harvard–Smithsonian Center of Astrophysics) and his colleagues have pointed out that the bar exerts a torque on gas beyond a certain radius, causing it to lose angular momentum. The gas flows inward and accumulates at 400 to 500 light years, leading to the formation of giant molecular clouds. These clouds migrate inward due to dynamical friction with the bulge stars; the resulting activity of the black hole and/or star formation blows the gas away and the process repeats.

Although these are useful speculations, it is presently unclear how activity of the black hole directly, or indirectly due to the possible of triggering star formation, affects the gas kinematics and distribution in the central region of the Galaxy. There is, however, increasing direct observational evidence for ongoing activity at a power level lower than the Eddington limit of the black hole.

12.2   X-Ray Echos

The gas in the central few hundred light years of the galaxies is concentrated in several massive molecular clouds such as Sagittarius B2, a cloud with a projected distance of about 390 light years from the center having a mass of three million solar masses; it is one of the most massive molecular clouds in the Milky Way. A remarkable discovery of the past twenty years is that, in X-rays, these clouds can act as gigantic mirrors, reflecting the variable emission from the accreting black hole associated with Sgr A*. And because the light travel time from Sgr A* to the Earth by way of the molecular cloud is longer than the direct straight line path from the Galactic Center (see Figure 12.2), it offers a fairly recent historical record of the activity of the black hole – how its power output, at least in X-rays, has varied over the past several hundred years.

The idea was first suggested in 1993 by Raschid Sunyaev (then at the Institute for Space Research in Moscow) and his colleagues. X-rays from the direction of the Galactic Center, at energies of several thousand of kilo-electron volts (keV), had been observed for some years. It was known that, in addition to the point-like source associated with Sgr A* itself (mentioned earlier), there was a larger source in the center extended about one degree along the plane of the Galaxy. Based on new (at that time) satellite observations covering an energy range up to 22 keV, Sunyaev and colleagues noticed that there was a possible association of X-ray emission with the molecular clouds such as Sgr B2. On this basis they suggested the mirror model. Because the energies of the X-rays are so large, they rejected the idea that this could be *in situ* emission by hot gas; the gas would have to be so hot that it could not be gravitationally confined to the cloud or the Galaxy. Thus, they reasoned, the X-rays must be reflected from another source.

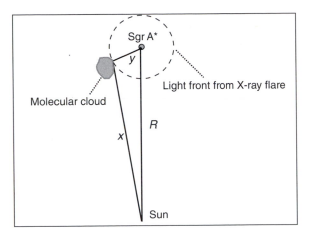

Figure 12.2 Geometry of X-ray flare reflection. A flare occurs at Sgr A* and the light front (dashed circle) propagates to the molecular cloud. The flare is then reflected in the direction of the Sun. Obviously the path of the reflected flare ($X + Y$) is longer than the direct path along R, so the reflected flare arrives at Earth several hundred years later than the direct flare from Sgr A*. As the light front propagates through the cloud the apparent velocity can be faster than the speed of light.

The mechanism of "reflection" is basically electron scattering. Free electrons scatter electromagnetic radiation of all energies equally – its cross section for scattering is the same from radio waves to gamma rays. In a molecular cloud most electrons are bound to atoms in molecules so they would hardly seem free. But to an X-ray of several keV, the binding energy of the electron is much smaller than the energy of the photon; most of the electrons appear to be free. So for X-ray photons a molecular cloud is a large bag of free electrons, available to scatter – reflect – the photons.

But there is more to it than that because X-ray spectral lines are actually observed from the clouds; primarily, a 6.4-keV line from iron. These lines are also scattered. This works because an X-ray with energy greater than 7.1 keV can partially ionize an iron atom by knocking out an electron – the most tightly bound electron in the iron atom. The ion then recombines with a free electron. The electron cascades down to the lowest energy state, emitting lines along the way, with the most energetic line being the final transition into the ground state; this is the 6.4 keV emission line. Basically this is the same fluorescence mechanism that works for the emission lines from regions of ionized hydrogen around hot young stars but, of course, translated to much higher energies.

So the molecular clouds can reflect the X-ray emission from a distant source, and we can observe a time-lapse picture of the emission history of the source.

The association of X-ray emission with individual molecular clouds became evident with higher resolution X-ray observations of the central region, in particular, those of the ASCA satellite reported by Katsuji Koyama and colleagues of Kyoto University. But the X-ray flux from Sgr B2 can be accounted for only in the context of this mechanism if the X-ray power of Sgr A* was as high as one million solar luminosities 300 years ago. This is direct evidence that Sgr A* was ten thousand times more luminous in the recent past than it is at present. This is still 100 000 times less than the Eddington limit, but it does imply a highly variable accretion rate onto the black hole.

This interpretation of the X-ray glow from molecular clouds has become quite definite with the discovery of flux and structure variations in the X-ray continuum structure of clouds. That is to say, the structure of the X-ray intensity is observed to vary in some molecular clouds in a manner that is consistent with the propagation of an X-ray light front through the cloud; in fact, the structure changes so rapidly that it would imply a superluminal motion (faster than the speed of light) if interpreted in terms of excitation by cosmic rays. It is clearly not a physical propagation but a motion of the reflection region.

It is difficult to disentangle the history of the flux variations from the actual distribution of the molecular clouds in space (we only observe their position projected onto the sky). But the implied power of these events, on the order of one million solar luminosities, is larger than would be expected from an event on the scale of a stellar mass – accretion onto a neutron star or stellar mass black hole. This is convincing evidence that the massive black hole is the source of the X-ray flares and that it has been considerably more active in the past several hundred years than at present.

12.3   Fermi Bubbles

In June 2008 NASA launched the Fermi Gamma-ray Space Telescope into low Earth orbit. The primary instrument on board is the Large Area Telescope (LAT), which covers about 20% of the sky with a resolution going down to a few arc minutes. The detectors are sensitive to very high energy photons, ranging from 30 MeV (million electron volts) up to 300 GeV (giga or billion electron volts). For comparison, the rest energy of the electron is 0.5 MeV and that of the proton is about 2 GeV, so the detected photons are energetic indeed. The project is truly an international effort, with the involvement of institutions in eight countries in addition to the United States.

So far, one of the more exciting discoveries of Fermi-LAT has been the detection of two enormous lobes of gamma ray emission above and below the plane of the Galaxy in the direction of the Galactic Center (see Figure 8.3). This structure,

Figure 12.3 A schematic diagram of the Fermi bubbles extending 25 000 light years above and below the plane of the Galaxy. These are the lobes of gamma ray emission detected by the Fermi Large Area Telescope. The reported jets, making an angle of about 15 degrees to the axis of the lobes, are also shown. (From Su and Finkbeiner 2012.) Credited to David A. Aguilar, courtesy of CfA Public Affairs Office.

discovered by Meng Su, Tracy Slater, and Douglas Finkbeiner in 2010, is reminiscent of the giant radio-emitting lobes surrounding radio galaxies; the energy of the individual photons, however, is sixteen factors of ten times larger. It is a large structure extending 25 000 light years above and below the plane of the Galaxy (see Figure 12.3).

These extremely high-energy photons, the gamma rays, are unquestionably produced by the scattering of background photons (radio to infrared) by very energetic electrons – the so-called inverse Compton mechanism. So the presence of the bubbles indicates a large population of highly relativistic electron cosmic rays. But if there are energetic electrons and a magnetic field, then would we also expect synchrotron radiation? Indeed there has been evidence for several years for this population of relativistic electrons from microwave observations. At tens of giga-Herz (tens of billion cycles per second corresponding to a wavelength of a few centimeters) the Wilkinson Microwave Anisotropy Probe (WMAP), the instrument designed to map the anisotropies in the primordial microwave background (the glow from the Big Bang), has observed a Galactic foreground called the WMAP haze. A number of explanations for the microwave haze have been proposed – spinning dust particles, thermal radiation from hot gas, decay of dark matter particles – but now the preferred explanation is that this mysterious source is synchrotron radiation from high-energy electrons that produce the gamma rays in the Fermi bubbles; the morphology of the WMAP haze is consistent with this interpretation.

So what is the explanation of the Fermi bubbles and what does it have to do with the black hole? In several nearby galaxies such as the "exploding" galaxy M82, similar structures are seen in the morphology of gas above and below the plane; these are the so-called starburst galaxies in which there appears to be rather recent and dramatic star formation activity in the plane of the galaxy itself. The newly formed stars, by means of their own winds or resulting supernovae, create substantial outflow of gas from the galaxy, which perhaps can convect the energetic particles out into the regions beyond the plane (see Veilleux et al. for a review). The morphology of these outflows is certainly similar to that of the Fermi bubbles. Perhaps, in the case of the Milky Way, we are seeing the effects of the starburst is triggered by activity of the Galactic Center black hole.

But a more direct connection is suggested by a recent result published by Su and Finkbeiner: they report the presence of a two-sided jet within the gamma ray–emitting lobes – a jet seen in the inverse Compton emission of gamma rays (see Figure 12.3). We know that in active galactic nuclei such as Seyferts and radio galaxies, accretion of matter in the form of a disk onto a black hole often, if not always, leads to the formation of a jet perpendicular to the disk axis. So it would be natural to assume that the same happens in the case of the Galactic Center black hole. The observation of Su and Finkbeiner would be the first direct evidence for a large-scale jet formed by an accretion event in the Galaxy. The total power of the jet, at one hundred times the luminosity of the Sun, is consistent with the present power of Sgr A*. Moreover the total energy in the Fermi bubbles, estimated to be on the order of 10 solar rest masses, is also consistent with the energy in the anomalous cloud motions in the plane of the Galaxy – the expanding molecular ring. So perhaps the Fermi bubbles originated in the previous major accretion event – an event that also caused a hydrodynamical disturbance of the gas in the Galactic plane.

12.4   A Snack for Pele

Human sacrifice to volcano gods is the stuff of Hollywood. This is the theme of an old film from 1952 starring Debra Pagent, Louis Jourdan, and Jeff Chandler. At the exciting conclusion pretty Debra, playing a native girl on a South Sea island paradise and beloved by Louis Jourdan, leaps into an active active volcano to appease the local god. It works. After a brief flare the volcano calms down, and the natives (and Louis) are saved. In reality, there is scant evidence for human sacrifice to volcanos, but there is evidence that a morsel is about to be sacrificed to the Pele at the heart of the Galaxy. And, as in the film, it is likely to make her flare briefly.

This evidence is the recent discovery of small cloud reported by Stefan Gillessen and the Max Planck group. The cloud, seen in an infrared hydrogen emission line, has an estimated mass of about three times that of Earth and lies on an orbit that will take it to about 3000 times the Schwarzschild radius. At that point the cloud should be tidally disrupted and increase the accretion rate onto the black hole; that is, the black hole is about to flare if this interpretation is correct (perhaps even before the publication of this book). The object is seen to have a finite size and is apparently being stretched along its orbit by the black hole tides.

This interpretation is not without problems, as pointed out by Mark Morris, Leo Meyer, and Andrea Ghez. A cloud with this mass would be gravitationally unbound and therefore it is curious that it exists at all. In an interpretative paper Andreas Burkert and colleagues address this issue; they propose that the cloud

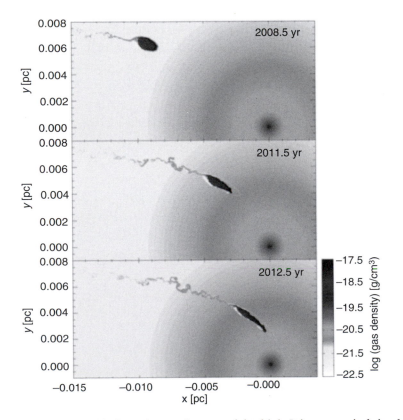

Figure 12.4 This figure is not what you might think. It is a numerical simulation of the small cloud on its way to the black hole, being stretched by tides. (From Burkert et al. 2012.)

was very recently formed due to the collisions of hot winds from the young stars and is on its first passage toward the center (see Figure 12.4). Alternatively, the cloud could be circumstellar, that is to say, associated with a young or evolved star: for example, a recent ejection of the envelope of a red giant star due to a collision with a normal dwarf star. There are several other suggested scenarios, but, in any case, the cloud is observed, it is extended in size, it is being tidally stretched, and in the next year or so, the fireworks should begin.

12.5   The Face of Pele: The Event Horizon Telescope

The Galactic Center black hole is the closest massive black hole. This means that the angular size of the Schwarzschild radius, the event horizon, appears larger to us than any other black hole in the Universe (except, perhaps, for the 10 billion solar mass black hole in the nucleus of M 87): the angular diameter should be on the order of 20 micro arc seconds (2×10^{-5} arc seconds). This corresponds quite closely to the resolution possible with Very Long Baseline Interferometry at millimeter or submillimeter wavelengths.

How would an emission region surrounding a black hole, an accretion region, appear to a distant observer? This has been worked out by Heino Falcke (Nijmegen), Fulvio Melia (Arizona), and Eric Agol (Johns Hopkins), among others. Light rays, or photons, are strongly deflected in the gravitational field of the black hole. Below a certain radius all of the emitted photons intersect the event horizon and are lost. Beyond that radius, the photons escape and can be observed, although they may orbit the black hole several times. This gives the appearance of bright ring surrounding a darker region – the "shadow" of the black hole, as in Figure 12.5. Due to the strong deflection of photons, this ring has a radius of about five Schwarzschild radii which, at the distance of Sgr A*, would correspond to about 50 micro arc seconds (5×10^{-5} arc seconds). Although these calculations assume that the accretion region is spherical cloud surrounding the black hole, the results for an accretion disk are qualitatively similar. The appearance of the hole shadow is also fairly independent of the rotation of the black hole or its viewing angle (although the symmetry of the image does change depending on these factors). And, for the Galactic Center black hole, this view of the event horizon is, in principle, observable using techniques of VLBI at millimeter wavelengths!

Apart from the high resolution available at shorter wavelengths observations of Sgr A* at wavelengths on the order of or less than 1 millimeter have the advantage that the source can be observed directly. At longer wavelengths scattering by turbulent cells in the intervening interstellar medium broaden and blur the source, and we cannot observe it directly. In 2007 Sgr A* was observed at a wavelength of 1.3 mm with VLBI by a large group headed by Sheperd Doeleman (MIT).

Figure 12.5 The shadow of the black hole. The dark area near the center is due to missing photons that have entered the event horizon. The ring is an enhancement of intensity due to photons that orbit the black hole before escaping. (From Falke, Melia, and Agol, 2000.)

There were three stations, in Arizona, California, and Mauna Kea in Hawaii. The source was seen with a size of about 37 micro-arc seconds, on the order of four times the size of the Schwarzschild radius.

Because there were only three stations, not much more can be said about the structure of the source than this estimate of its size, but the fact that there is structure on the scale of the event horizon is very encouraging. It has encouraged the proposal of a larger VLBI network to image Sgr A* on the scale of the horizon – the Event Horizon Telescope. This would be, initially, a six-station array scattered around the globe of the earth. One station would be the new submillimeter ALMA array – the Atacama Large Millimeter Array located, of course, in the Atacama desert in northern Chile. So once again these two legendary locations, Mauna Kea and Atacama, will contribute to the understanding of, or perhaps even confirm, the presence of this strange object surrounded by an event horizon in the center of the Galaxy.

13

After Words: Progress in Astronomy

It is remarkable that a mere one hundred years ago most astronomers thought that the Universe was the Galaxy, a large flattened system of stars with the Sun near the center and various fuzzy objects scattered around it. This was the view supported by the systematic study of the positions and motions of stars began here in Groningen by Jacobus Kapteyn, a view that turned out to be wrong because no correction was made, or could made at that time, for the obscuration of star light by interstellar dust particles. The true scale of the Milky Way and the position of the Sun in the system became evident only after the pioneering work by Harlow Shapley on the spacial distribution of globular clusters in the halo of the Galaxy, work in turn based on the fundamental discovery of the period-luminosity relation for Cepheid variable stars by Henrietta Leavitt.

But Kapteyn's precise measurements of positions and motions of stars turned out to be extremely useful for his student, Jan Oort, who used the data to support the idea of Galactic rotation and accurately described the scale of the Milky Way and the Sun's true place, not in the center but on the outskirts of this vast system of stars. Oort, more than anyone, discovered the Center of the Galaxy – its direction and distance. Through radio astronomy and the 21-cm line of neutral hydrogen he and his students discovered the center of rotation, that there were expanding gas features moving away from that center, and that the center coincided with the source of continuum radio waves, the brightest radio source in the constellation of Sagittarius.

It was not even one hundred years ago when the mathematical basis for black holes, Einstein's theory of gravity, was formulated. Black holes were implicit in Karl Schwarzschild's solution of Einstein's equations for the gravitational field about a spherical object, but this was not appreciated for some years, one might say not until 1930 and Subrahmanyan Chandrasekhar's discovery that stable white dwarf stars were not possible with mass higher than one and one-half solar masses. More massive stars that had exhausted their hydrogen fuel could not support themselves against gravity. Even so, the objects permitted by Schwarzschild's solution, with horizons and singularities, were considered entirely hypothetical, and the majority of astrophysicists thought that nature would contrive to avoid such strange constructs. The discovery in 1963 of quasi-stellar radio sources, quasars, with extreme energy emerging from a very compact region, pushed such objects to the forefront, but this time as extremely massive concentrations in galactic nuclei.

Donald Lynden-Bell made the step toward uniting the concept of a massive black hole with the Milky Way Center and normal galaxies in general – as remnants of an earlier quasar period. This was the birth of the current paradigm of active galactic nuclei that has become almost universally accepted. It is a fantastic story; science fiction could not be more exciting.

In this discussion of the discovery of the Milky Way and its massive black hole I have, at several points, drawn general conclusions about how the science of astronomy advances – how new ideas emerge and how new models are formulated. The development of the black hole paradigm of active galactic nuclei provides a particularly relevant illustration of these generalities. Astronomy advances primarily through observations and not theoretical speculations. The subject is driven forward by observations. Theories are primarily developed to explain existing observations and the often serendipitous discoveries made through observations. Occasionally theoretical models do make predictions that are brilliantly correct, for example, the prediction by Martin Rees of superluminal motions in compact radio sources before these were actually found. But then these were not discovered because of Rees' prediction; it was a discovery made as part of an ongoing research program in Very Long Baseline Interferometer – a discovery that happened to come after the prediction. This does not detract at all from Rees' achievement; the explanation for this phenomenon was waiting in the pages of *Nature*.

Those observations that lead to new insights result from new technology that opens a new wavelength band, and this new technology has most often developed due to urgent demands of national defense. Quasars were discovered as a result of the advent of radio astronomy that followed directly from the development of radar in World War II. These extreme compact objects, with a total power that seemed to stretch the possibilities of known physics, led to a plethora of models, only one of which survived – the black hole model developed primarily by Donald Lynden-Bell and Martin Rees at Cambridge. But when proposed, this seemed no more likely than several of the competitors. The process of extensive construction of models and then the narrowing of possibilities has been repeated over the past half-century. So it was with quasars and so it was with pulsars that came fifteen years later. So it has been with gamma ray burst sources that have, only relatively recently, been identified with massive explosions of stellar mass objects in distant galaxies.

It is the new technology that has driven this process. The development of adaptive optics also had a military motivation – to remove the blurring of the atmosphere in identifying and observing the satellites of a perceived enemy. But when applied on 10-meter class telescopes this led to the definitive determination of the mass and the limit on the size of the central concentration in the Galactic nucleus and to the conclusion that it could only be a black hole.

The developments from radio interferometry to adaptive optics were made possible by the simultaneous exponential growth of machine computing power. To adjust the flexible mirror along the optical path of a 10-meter telescope many times per second in order to correct for atmospheric blurring would have been unthinkable without high-speed computers. The computer has become an essential tool for real-time astronomical observations, not just post-observation analysis.

Of course, theory and model-building play a vital role in this process. It is necessary to have a range of models to provide a framework for observation – to suggest new observations. And most of these models will be wrong. In astronomy, a theoretician does not have to be right to contribute to progress; it is important to be provocative.

Even the right model will not be entirely right. Lynden-Bell did not imagine the black hole in the Milky Way exactly as it turned out to be. He perceived a very massive object, 100 million solar masses, continuously accumulating matter at a declining rate through the giant accretion disk of the Galaxy spiraling down into the hole-like water in the drain of a bath tub. Through the hindsight provided by forty years of observations we now have a more accurate idea of the scale of the object (only 4.5 million solar masses), the highly stochastic nature of accretion events (with distinct episodes on timescales of hundreds to millions of years), and the accompanying bursts of star formation.

A range of theories, or models – even apparently outrageous models – sets the stage for the most important element of progress: the conflict between ideas. In 1981 there was a meeting on extragalactic radio sources in Albuquerque, New Mexico, celebrating the first few years of operation of the Very Large Array in the New Mexico desert and the initial results. There was still argument over the energy source of radio galaxies. Some believed in the role of massive explosions and ejections, and this viewpoint did not appear as unreasonable then as it does now. After all, gas can accumulate in the nuclei of galaxies, form a massive star or rotating object that is unstable, and this instability can lead to a break up and ejection. Which is more reasonable – such a scenario involving conceivable astrophysical objects or a massive dark object with an event horizon hiding a central mathematical singularity? The black hole model won because this state, in a real sense, is inevitable (in the context of known physics) and long-lived and because it provides the most efficient means of converting mass into observable energy. It is most consistent with the gradual, not explosive, input of relativistic particles into the giant radio emitting lobes via jets – jets that by that point were actually observed. But an important aspect of formulating these arguments was the existence of alternatives.

In establishing the black hole model, most significant were the actual observations of star orbits very near the center of the Galaxy – observations by the groups of Reinhard Genzel and Andrea Ghez that left no viable alternative to the black hole models. The activities of these two groups so closely track one another that it is not possible to assign priority. Both groups have, beyond reasonable doubt, proven the presence of massive black hole in the Galactic Center.

I have mentioned at several points that in explaining new phenomena it is wiser for theoretical astrophysicists to stick with known physics rather than venture into the speculative wilderness of new and unspecified physics. I believe this to be true, in general. However, we should recall that massive black holes were certainly not conventional in 1969; although the concept has a venerable history going back to the early days of General Relativity, most astronomers and physicists did not imagine that such constructs could actually exit as real astrophysical objects. The ultimate criterion for viability of an idea, no matter how nonstandard it might seem, should be success in explaining the phenomena, and the black hole model has certainly met that criterion.

The idea of explosive ejections from galactic nuclei, advocated in extreme form by Ambartsumian and taken seriously by respectable people such as Oort, has turned out to be a diversion. But was it really a useless exercise? I have argued that attempting to model features such as the 3 kiloparsec expanding arm in the Milky Way or the anomalous spiral arms in NGC 4258 in terms of explosive ejections was indeed useful because it defined the extreme properties that such events must have: an explosion in the Galactic Center involving the ejection of one hundred million solar masses with a kinetic energy of ten thousand solar rest masses certainly does appear to be more drastic than steady gas flow on noncircular orbits in the nonaxisymmetric gravitational field of a bar – structures that we observe directly in the Milky Way as well as in other galaxies. And steady flow of matter in jets from a black hole does appear to be more plausible than cold clouds shot out as if from cannons from a mysterious central source. So such modeling is useful in that it sets the terms of plausibility arguments; however, it is a good idea for theoreticians not to fall too much in love with their own more extreme ideas and give them up (as did Oort) when the observational evidence becomes overwhelming.

But we have seen that even one of the more extreme proposals of Ambartsumian – that the nuclei of galaxies have a dramatic effect on the structure and evolution of the surrounding system – has returned in a new guise called "feedback." In some sense this is a final round of the process: proposals of alternative ideas, modeling and observations based on those ideas, conflict, resolution, and synthesis.

Progress in astronomy is a dialectic and a social process. The social aspect is, if anything, more prominent at present than in the past because the nature of observations (and computations) has changed. Large and expensive telescopes, space telescopes, long-baseline interferometry, multinational projects all mean that the number of astronomers involved in major observational programs has increased considerably as has the number of co-authors on papers reporting results. This has positive and negative aspects. More members on an observing campaign can mean more input of ideas, although it is also more difficult to identify the co-author responsible for a particular idea. I find it personally sad that the days of the single astronomer alone with the cold sky and only the telescope in between have also largely vanished (real observers might not find this so unfortunate). It is also the case that the individual theoretician with pen and paper is less prominent; theoretical work in astronomy has also come to rely on computing at large facilities with many co-workers. It is unclear if this development will have an entirely beneficial effect on the creative process in science.

But on the positive side, the family of astronomy has become far more inclusive than when I was young – in particular with respect to the number of women now involved in active research. It is a far cry from the time, one hundred years ago, when Mt. Wilson Observatory built a special accommodation for Kapteyn because he brought his wife from Groningen (the "Kapteyn cottage"). The presence of a larger fraction of women involved in research has changed the social mood strongly toward a more cooperative, less fiercely competitive, less testosterone-driven atmosphere – in observatories and institutes and at scientific meetings. Gone (hopefully) are the days when dog-eared old Playboys littered observatory control rooms and dormatories.

There is another very positive aspect of a more inclusive community and that has to do with the realm of ideas. Science in general, and astronomy in particular, aims at objectivity but in fact, because we are human, falls short of that aim. There are fashions in science as in every other aspect of human activity, and these fashions are evident in the problems considered important and in the vocabulary used. For example, galaxies no longer "form" but they are "assembled." This reflects the current view that galaxies build up over cosmological timescales by a series of mergers, rather than forming in a single monolithic collapse – that the process is hierarchical. There is a great deal to be said for this viewpoint, but the language used does tend to predetermine the conclusions drawn. The same is true of mass discrepancies in galaxies. It is considered a "dark matter problem" and not a "gravity problem." This use of fashionable but prejudicial language in science is in some sense unavoidable but could be alleviated by a more pluralistic community.

Apart from fashion, theories in science are more or less "underdetermined" – there is an inevitable incompleteness of the evidence available to fully justify scientific theories. There is a gap between evidence and justification (more apparent in the "soft" sciences but also true in physics and astronomy). As is argued by some philosophers of science, such as Helen Longino, that gap is filled by implicit assumptions and preconceptions that are intrinsic to the social background of those involved – determined by class, by race, by gender. To me this seems credible, and, if so, then a more heterogeneous community cannot but help to broaden the consideration of what is reasonable and lead to a fuller discussion of the possibilities. Not to mention considerations of fairness, this can only be positive for progress.

But I believe that the advantages of diversity in scientific communities goes beyond the issue, primarily of interest to philosophers of science, of underdetermination of theories. Such diversity is also good for experimental efforts and observational campaigns. Again the competition between the European and American groups on stellar orbits in the Galactic Center is a perfect illustration of this. Although they generally treated each other with politeness and respect, there was, and is, certainly tension between the two groups – a tension resulting from the simple desire to be first that is exacerbated by differences in culture and approach. Yet it is this very tension that drives both groups forward. The fact that there is a competitor ready to pounce on any mistake or error makes a scientist very careful about announcing results or making claims. The fact that the two diverse groups reach the same conclusions provides an overall confidence in these results. Although stressful at times, such tension is generally beneficial for the scientific process.

My personal journey to the center of the Galaxy has led me from a boyhood under the brilliant Texas sky to the halls of academia to the damp polders of the north of the Netherlands. How on earth did I wind up here so far away from my native land? In my decision to live my life away from home and country there were, of course, personal considerations as well as professional. But above all, it was the warm and friendly atmosphere of the Kapteyn Institute – this very important social aspect of the practice of science – that induced me to stay. When

I first came here thirty-five years ago, the Westerbork interferometer was new and producing exciting results every day, or so it seemed. The collection of scientists here was international and united by an effort to get the most out of this forefront instrument, but in a quite collegial way with free discussion of results and ideas. This made a profound impression on me.

But now, as I become older my dreams return to those of a small boy growing up in Texas and marveling at the summer night sky with its luminous band of the Milky Way. This I believe will always remain – a prescientific wonder at the Universe and the mystery of our place in it. And through it all runs the Milky Way. The Milky Way – sky river, Ganges of heaven, pathway of the gods.

References

Allen, D.A., Hyland, A.R., and Hillier, D.H. (1990). The source of luminosity at the Galactic Centre. *Mon. Not. RAS*, **244**, 706–713.

Allen, D.A., and Sanders, R.H. (1986). Is the Galactic Centre black hole a dwarf? *Nature*, **319**, 191–194.

Ambartsumian, B.A. (1976). The role of nuclear activity in the overall evolutionary processes in galaxies. *Proc. 3rd Europ. Astron. Meeting* (Tiflis), 91–96.

Babcock, H.W. (1953). The possiblility of compensating astronomical seeing. *Pub. Astron. Soc. Pacific*, **65**, 229–236.

Backer, D.C., and Sramek, R.A. (1982). Apparent proper motion of the Galactic Center compact radio source and PSR 1929+10. *Astrophys. J.*, **260**, 512–519.

Baganoff, F.K., Bautz, M.W., Brandt, W.N., Chartas, G., Feigelson, E.D., Garmire, G.P., Maeda, Y., Morris, M., Ricker, G.R., Townsley, L.K., and Walter, F. (2001). Rapid X-ray flaring from the direction of the supermassive black hole at the Galactic Centre. *Nature*, **413**, 45–48.

Balick, B., and Brown, R.L. (1974). Intense sub-arcsecond structure in the Galactic Center. *Astrophys. J.*, **194**, 265–270

Barthel, P.D. (1989). Is every quasar beamed? *Astrophys. J.*, **336**, 606–611.

Becklin, E.E., and Neugebauer, G. (1968). Infrared observations of the Galactic Center. *Astrophys. J.*, **151**, 145–161.

Becklin, E.E., and Neugebauer (1975). High-resolution maps of the Galactic Center at 2.2 and 10 microns. *Astrophys. J.*, **200**, L71–L74.

Becklin, E.E., Gatley, I., and Werner, M.W. (1982). Far-infrared observations of Sagittarius A: The luminosity and dust density in the central parsec of the Galaxy. *Astrophys. J.*, **258**, 135–142.

Begelman, M.C., Volonteri, M., and Rees, M.J. (2006). Formation of super-massive black holes by direct collapse in pre-galactic haloes. *Mon. Not. RAS*, **370**, 289–298.

Binney, J., Gerhard, O.E., Stark, A.A., Bally, J., and Uchida, K. (1991). Understanding the kinematics of the Galactic Centre gas. *Mon. Not. RAS*, **252**, 210–218.

Blandford, R.D., and Rees, M.J. (1974). A 'twin-exhaust' model for double radio sources. *Mon. Not. RAS*, **169**, 395–415.

Blandford, R.D., and Znajek, R.L. (1977). Electromagnetic extraction of energy from Kerr black holes. *Mon. Not. RAS*, **179**, 433–456.

Blitz, L., and Spergel, D.N. (1991). Direct evidence for a bar at the Galactic Center. *Astrophys. J.*, **379**, 631–638.

Burbidge, G.R. (1959). Estimates of the total energy in particles and magnetic field in the non-thermal radio sources. *Astrophys. J.*, **129**, 849–851.

Burbidge, G.R., Burbidge, E.M., and Sandage, A.R. (1963). Evidence for the occurrence of violent events in the nuclei of galaxies. *Rev. Mod. Phys.*, **35**, 947–980.

Burkert, A., Schartmann, M., Alig, C., Gillessen, S., Genzel, R., Fritz, T.K., and Eisenhauer, F. (2012). Physics of the Galactic Center cloud G2 on its way toward the supermassive black hole. *Astrophys. J.*, **750:58**, 17pp.

Burton, M., and Allen, D.A. (1992). Imaging the hot molecular gas at the Centre of the Galaxy. *Proc. Astronom. Soc. Aus.*, **10**, 55–57.

Burton, W.B. (1972). On the kinematic distribution of Galactic neutral hydrogen. *Astron. Astrophys.*, **19**, 51–65.

Chandrasekhar, S. (1931). The maximum mass of ideal white dwarfs. *Astrophys. J.*, **74**, 81–82.

Christiansen, W.N., and Hindman, J.V. (1952). 21 cm line radiation from galactic hydrogen. *The Observatory*, **72**, 149–151.

Colgate, S.A. (1967). Stellar coalescence and the multiple supernova interpretation of quasi-stellar sources. *Astrophys. J.*, **150**, 163–192.

Collins, M. (2005). Pele and Polianhu: A tale of fire and ice. BeachHouse Publishing.

Croton, D.J., Springel, V., White, S.D.M., De Lucia, G., Frenk, C.S., Gao, L., Jenkins, A., Kauffmann, G., Navarro, J.F., and Yoshida, N. (2006). The many lives of active galactic nuclei: Cooling flows, black holes and the luminosities and colours of galaxies. *Mon. Not. RAS*, **365**, 11–28.

de Vaucouleurs, G. (1964). Interpretation of velocity distribution of the inner regions of the Galaxy. In *I.A.U. Symp.20, The Galaxy and Magellanic Clouds*, (Canberra), ed. F.J. Kerr, 195–199.

Doeleman, S.S., et al. (2008). Event-horizon-scale structure in the supermassive black hole candidate at the Galactic Centre. *Nature*, **455**, 78–80

Dressler, A., and Richstone, D. (1988). Stellar dynamics in the nuclei of M31 and M32 – Evidence for massive black holes? *Astrophys. J.* **324**, 701–713.

Eckart, A., and Genzel, R. (1996). Observations of proper motions near the Galactic Center. *Nature*, **383**, 415–417.

Eisenhauer, F., Genzel, R., Alexander, T., Abuter, R., Paumard, T., Ott, T., Gilbert, A., Gillessen, S., Horrobin, M., Trippe, S., Bonnet, H., Dumas, C., Hubin, N., Kaufer, A., Kissler-Patig, M., Monnet, G., Ströbele, S., Szeifert, T., Eckart, A., Schödel, R., and Zucker, S. (2005). SINFONI in the Galactic Center: Young stars and infrared flares in the central light month. *Astrophys. J.*, **628**, 246–259.

Ekers, R.D., and Lynden-Bell, D. (1971). High resolution observations of the Galactic Center at 5 GHz. *Astrophys. Lett.*, **9**, 189–193.

Ekers, R.D., Goss, W.M., Schwarz, U.J., Downes, D., and Rogstad, D.H. (1975). A full synthesis map of Sgr A at 5 GHz. *Astron. Astrophys.*, **43**, 159–166.

Ekers, R.D., van Gorkom, J.H., Schwarz, U.J., and Goss, W.M. (1983). The radio structure of Sgr A. *Astron. Astrophys.*, **122**, 143–150.

Emonts, B.H.C., Morganti, R., Tadhunter, C.N., Oosterloo, T.A., Holt, J., and van der Hulst, J.M. (2005). A jet-induced outflow of warm gas in 3C293. *Mon. Not. RAS*, **362**, 931–944.

Ewen, H.I., and Purcell, E.M. (1951). Observation of a line in the Galactic radio spectrum: Radiation from Galactic hydrogen at 1420 Mc/sec. *Nature*, **168**, 356–359.

Fabian, A.C., Wilman, R.J., and Crawford, C.S. (2002). On the detectability of distant Compton-thick obscured quasars. *Mon. Not. RAS*, **329**, L18–L22.

Falcke, H., Melia, F., and Agol, E. (2000). Viewing the shadow of the black hole at the Galactic Center. *Astrophys. J.*, **528**, L13–L16.

Feast, M., and Whitelock, P. (1997). Galactic kinematics of Cepheids from Hipparcos proper motions. *Mon. Not. RAS*, **291**, 683–693.

Ferrarese, L., and Merritt, D. (2000). A fundamental relation between supermassive black holes and their host galaxies. *Astrophys. J.*, **519**, L9–L12.

Field, G.B. (1964). Quasi-stellar radio sources as spherical galaxies in the process of formation. *Astrophys. J.*, **140**, 1434–1444.

Foy, R., and Labeyrie, A. (1985). Feasibility of adaptive telescope with laser probe. *Astron. Astrophys.*, **152**, L29–L31.

Gebhardt, K., Bender, R., Bower, G., Dressler, A., Faber, S.M., Filippenko, A.V., Green, R., Grillmair, C., Ho, L.C., Kormendy, J., Lauer, T.R., Magorrian, J., Pinkney, J., Richstone, D., and Tremaine, S. (2000). A relationship between nuclear black hole mass and galaxy velocity dispersion. *Astrophys. J.*, **539**, L13–L16.

Genzel, R., Schödel, R., Ott, T., Eisenhauer, F., Hofmann, R., Lehnert, M., Eckart, A., Alexander, T., Sternberg, A., Lenzen, R., Clénet, Y., Lacombe, F., Rouan, D., Renzini, A., and Tacconi-Garman, L.E. (2003). The stellar cusp around the supermassive black hole in the Galactic Center. *Astrophys. J.*, **594**, 812–832.

Genzel, R., Schödel, Ott, T., Eckart, A., Alexander, T., Lacombe, F., Rouan, D., and Aschenbach, B. (2003). Near-infrared falres from accreting gas around the supermassive black hole at the Galactic Centre. *Nature*, **425**, 934–937.

Genzel, R., Thatte, N., Krabbe, A., Kroker, H., and Tacconi-Garman, L.E. (1996). The dark mass concentration in the central parsec of the Milky Way. *Astrophys. J.*, **472**, 152–173.

Ghez, A.M., Duchêne, G., Matthews, K., Hornstein, S.D., Tanner, A., Larkin, J., Morris, M., Becklin, E.E., Salim, S., Kremenek, T., Thompson, D., Soifer, B.T., Neugebauer, G., and McLean, I. (2012). An ultraviolet-optical flare from the tidal disruption of a helium-fich stellar core. *Nature*, **485**, 217–220.

Ghez, A.M., Duchêne, G., Mathews, K., et al. (2003). The first measurement of spectral lines in a short-period star bound to the Galaxy's central black hole: A paradox of youth. *Astrophys. J.*, **586**, L127–L131

Ghez, A.M., Morris, M.R., Becklin, E.E., Tanner, A., and Kremenek, T. (2000). The accelerations of stars orbiting the Milky Way's central black hole. *Nature*, **407**, 349–351.

Ghez, A.M., Salim, S., Hornstein, S.D., Tanner, A., Lu, J.R., Morris, M., Becklin, E.E., and Duchne, G. (2005). Stellar orbits around the Galactic Center black hole. *Astrophys. J.*, **620**, 744–757.

Ghez, A.M., Salim, S., Weinberg, N.N., Lu, J.R., Do, T., Dunn, J.K., Mathews, K., Morris, M.R., Yelda, S., Becklin, E.E., Kremenek, T., Milosavljevc, M., and Naiman, J. (2008). Measuring distance and properties of the Milky Way's central supermassive black hole with stellar orbits. *Astrophys. J.*, **689**, 1044–1062.

Gillessen, S., Eisenhauer, F., Trippe, S., Genzel, R., Martins, F., and Ott, T. (2009). Monitoring stellar orbits around the massive black hole in the Galactic Center. *Astrophys. J.*, **692**, 1075–1109.

Greenstein, J.L., and Schmidt, M. (1964). The quasi-stellar radio sources 3C 48 and 3C 273. *Astrophys. J.*, **140**, 1–34.

Gültekin, K., Richstone, D.O., Gebhardt, K., Lauer, T.R., Tremaine, S., Aller, M. C., Bender, R., Dressler, A., Faber, S.M., Filippenko, A.V., Green, R., Ho, L.C., Kormendy, J., Magorrian, J., Pinkney, J., and Siopis, C. (2009). The $M - \sigma$ and $M - L$ relations in galactic bulges and determination of their intrinsic scatter. *Astrophys. J.*, **698**, 198–221.

Güsten, R., Genzel, R., Wright, M.C.H., Jaffe, D.T., Stutzki, J., and Harris, A.I. (1987). Aperture synthesis observations of the circumnuclear ring in the Galactic Center. *Astrophys. J.*, **318**, 124–138.

Hazard, C., Mackey, M.B., and Shimmins, A.J. (1963). Investigation of the radio source 3C 273 by the method of lunar occultations. *Nature*, **197**, 1037–1038.

Hills, J.G. (1975). Possible power source of Seyfert galaxies and QSOs. *Nature*, **254**, 295–298.

Hills, J.G. (1988). Hyper-velocity and tidal stars from binaries disrupted by a massive Galactic black hole. *Nature*, **331**, 687–689.

Hoffmann, W.F., Frederick, C.F., and Emerey, R.J. (1971). 100-micron map of the Galactic Center region. *Astrophys. J.*, **164**, L23–L28.

Hoyle, F., and Fowler, W.A. (1963). On the nature of strong radio sources. *Mon. Not. RAS*, **125**, 169–176.

Jahnke, K., and Macció, A. (2011). The non-causal origin of the black hole-galaxy scaling relations. *Astrophys. J.*, **734:92**, 11 pp.

Kaifu, N., Kato, T., and Iguchi, T. (1972). 270 pc expanding ring at the Galactic Center. *Nature Phys. Sci*, **238**, 105–107.

Keller, C. (1972). Mitos y Leyendas de Chile, Enciclopedia Moderna de Chile

Kerr, R.P. (1963). Gravitational field of a spinning mass as an example of algebraically special metrics. *Phys. Rev. Lett.*, **11**, 237–238.

Krul, W. (2000) "Kapteyn and Groningen: a Portrait" in *The Legacy of J.C. Kapteyn* eds. P.C. van der Kruit and K. van Berkel, Kluwer, Dordrecht, the Netherlands, pp. 53–78.

Kormendy, J. (1988). Evidence for a supermassive black hole in the nucleus of M31. *Astrophys. J.*, **325**, 128–141.

Kormendy, J., and Richstone, D. (1995). Inward bound – the search for supermassive black holes in galactic nuclei. *Annu. Rev. Astron. Astrophys.*, **33**, 581–624.

Koyama, K. Maeda, Y., Sonobe, T., Takeshima, T., Tanaka, Y., and Yamauchi, S. (1996). ASCA view of our Galactic Center: Remains of past activities in X-rays? *Publ. Astron. Soc. Japan*, **48**, 249–255.

Kwee, K.K., Muller, C.A., and Westerhout, G. (1954). The rotation of the inner parts of the Galactic System. *Bull. Astron. Inst. Netherlands*, **12**, 211–222.

Lauer, T.R., Faber, S.M., Groth, E.J., Shaya, E.J., Campbell, B., Code, A., Currie, D.G., Baum, W.A., Ewald, S.P., Hester, J.J., Holtzman, J.A., Kristian, J., Light, R.M., Ligynds, C.R., O'Neil, E.J., Jr., and Westphal, J.A. (1993). Planetary camera observations of the double nucleus of M31. *Astron. J.*, **106**, 1436–1447.

Leavitt, H.S., and Pickering, E.C. (1912). Periods of 25 variable stars in the Small Magellanic Cloud. *Harvard Coll. Circ.*, **173**, 1–3.

Levin, Y., and Beloborodov, A.M. (2003). Stellar disk in the Galactic Center: A remnant of a dense accretion disk?. *Astrophys. J.*, **590**, L33–L36.

Liebling, S.L., and Panenzuela, C. (2012). Dynamical boson stars. *Liv. Revs. Rel*, **15**, no. 6

Light, E.S., Danielson, R.E., and Schwarzschild, M. (1974). The nucleus of M31. *Astrophys. J.*, **194**, 257–263.

Liszt, H.S., and Burton, W.B. (1980). The gas distribution in the central region of the Galaxy. III – A barlike model of the inner-Galaxy gas based on improved H I data. *Astrophys. J.*, **236**, 779–797.

Liszt, H.S., Burton, W.B., and van der Hulst, J.M. (1985). Associations between neutral and ionized gas in SGR A. *Astron. Astrophys.*, **142**, 237–244.

Liszt, H.S., van der Hulst, J.M., Burton, W.B., and Ondrechen, M.P. (1983). VLA synthesis of H I absorption toward SGR A. *Astron. Astrophys.*, **126**, 341–351.

Lu, J.R., Ghez, A.M., Hornstein, S.D., Morris, M.R., Becklin, E.E., and Mathews, K. (2009). A disk of young stars at the Galactic Center as determined by individual stellar orbits. *Astrophys. J.*, **690**, 1463–1487.

Lynden-Bell, D. (1969). Galactic nuclei as collapsed old quasars. *Nature*, **223**, 690–694.

Lynds, C.R., and Sandage, A.R. (1963). Evidence for an explosion in the center of the Galaxy M82. *Astrophys. J.*, **137**, 1005–1021.

Magorrian, J., Tremaine, S., Richstone, D., Bender, R., Bower, G., Dressler, A., Faber, S.M., Gebhardt, K., Green, R., Grillmair, C., Kormendy, J., and Lauer, T. (1998). The demography of massive dark objects in galaxy centers. *Astron. J.*, **115**, 2285–2305.

Maiolino, R., Gallerani, S., Neri, R., Cicone, C., Ferrara, A., Genzel, R., Lutz, D., Sturm, E., Tacconi, L.J., Walter, F., Feruglio, C., Fiore, F., and Piconcelli, E. (2012). Evidence of strong quasar feedback in the early Universe. *Mon. Not. RAS*, **425**, L66–L70.

Mathews, T.A., and Sandage, A.R. (1963). Optical identification of 3C 48, 3C 196 and 3C 286 with stellar objects. *Astrophys. J.*, **138**, 30–56.

Menten, K.M., Reid, M.J., Eckart, A., and Genzel, R. (1997). The position of Sagittarius A*: Accurate alignment of the radio and infrared reference frames at the Galactic Center. *Astrophys. J.*, **475**, L111–L114.

Miyoshi, M., Moran, J., Herrnstein, J., Greenhill, L., Nakal, N., Diamond, P., and Makoto, I. (1995). Evidence for a black hole from high rotation velocities in a sub-parsec region of NGC 4258. *Nature*, **373**, 127–129.

Morris, M., Ghez, A.M., and Becklin, E.E. (1999). The Galactic Center black hole: Clues for the evolution of black holes in galactic nuclei. *Adv. Space Res.*, **23**, 959–968.

Morris, M., Meyer, L., and Ghez, A.M. (2012). Galactic Center research: Manifestations of the central black hole. *Res. Astron. Astrophys.*, **12**, 995–1020.

Mulder, W.A., and Liem, B.T. (1986). Construction of a global gas-dynamical model for our galaxy. *Astron. Astrophys.*, **157**, 148–158.

Oort, J.H. (1927). Observational evidence confirming Lindblad's hypothesis of a rotation of the Galactic System. *Bull. Astron. Inst. Neth.*, **3**, 275–282.

Oort, J.H., and Muller, C.A. (1952). Observation of a line in the Galactic radio spectrum: The interstellar hydrogen Line at 1,420 Mc./sec., and an Estimate of galactic rotation. *Nature*, **168**, 357–358.

Oort, J.H., Kerr, F.J., and Westerhout, G. (1958). The galactic system as a spiral nebula. *Mon. Not. RAS*, **118**, 379–389.

Oort, J.H., and Rougoor, G.W. (1960). The position of the galactic centre. *Mon. Not. RAS*, 121, 171–173.

Oort, J.H. (1977). The Galactic Center. *Ann. Rev. Astron. Astrophys.*, **15**, 295–362.

Oort, J.H. (1985). The Galactic Nucleus. *The Milky Way Galaxy, Proc. IAU Symp. 106*, eds. H. van Woerden, R.H. Allen, and W.B. Burton, Dordrecht, The Netherlands: Reidel, 363–365.

Oppenheimer, J.R., and Volkoff, G.M. (1939). On massive neutron cores. *Phys. Rev.*, **55**, 374–378.

Perley, R.A., Dreher, J.W., and Cowan, J.J. (1984). The jet and filaments in Cygnus A. *Astrophys. J.*, **285**, L35–L38.

Phinney, E.S. (1989). Manifestations of a massive black hole in the Galactic Center. In *IAU Symp. 136, The Center of the Galaxy*, ed. Mark Morris, Dordrecht, The Netherlands: Kluwer, 543–553.

Readhead, A.C.S., Cohen, M.H., Pearson, T.H., and Wilkinson, P.N. (1978). Bent beams and the overall size of extragalactic radio sources. *Nature*, **276**, 768–771.

Rees, M.J. (1966). Appearance of relativistically expanding radio sources. **211**, 468–470.

Rees, M.J. (1984). Black hole models for active galactic nuclei. *Ann. Rev. Astron. Astrophys.*, **22**, 471–506.

Reid, M.J., Readhead, A.C.S., Vermeulen, R.C., and Treuhaft, R.N. (1999). The proper motion of Sgr A*. I. First VLBA results. *Astrophys. J.*, **524**, 816–823.

Rieke, G.H., and Low, F.J. (1973). Infrared maps of the Galactic Nucleus. *Astrophys. J.*, **184**, 415–425.

Rieke, G.H., and Lebofsky, M.J. (1982). Comparison of Galactic Center with other galaxies. *AIP Conf. Proc.*, **83**, 194–203.

Rougoor, G.W., and Oort, J.H. (1960). Distribution and motion of interstellar hydrogen in the galactic system with particular reference to the region within 3 kiloparsecs of the Center. *Proc. Natl. Acad. Sci. USA*, **46**, 1–13.

Ryle, M., Elsmore, B., and Neville, A.C. (1965). High resolution observations of the radio sources in Cygnus and Cassiopeia. *Nature*, **205**, 1259–1262.

Salpeter, E.E. (1964). Accretion of interstellar matter by massive objects. *Astrophys. J.*, **140**, 796–800.

Sanders, R.H. (1970). The effects of stellar collisions in dense stellar systems. *Astrophys. J.*, **162**, 791–809.

Sanders, R.H. (1998). The circumnuclear material in the Galactic Centre – A clue to the accretion process. *Mon. Not. RAS*, **294**, 35–46.

Sanders, R.H., and Prendergast, K.H. (1974). The possible relationship of the 3-kiloparsec arm to explosions in the Galactic Nucleus. *Astrophys. J.*, **188**, 489–500.

Saslaw, W.C., Valtonen, M.J., and Aarseth, S.J. (1974). The gravitational slingshot and the structure of extragalactic radio sources. *Astrophys. J.*, **190**, 253–270.

Scheuer, P.A.G. (1974). Models of extragalactic radio sources with a continuous energy supply from a central object. *Mon. Not. RAS*, **166**, 513–528.

Schödel, R., Ott, T., Genzel, R., Hofmann, R., Lehnert, M., Eckart, A., Mouawad, N., Alexander, T., Reid, M.J., Lenzen, R., Hartung, M., Lacombe, F., Rouan, D., Gendron, E., Rousset, G., Lagrange, A.-M., Brandner, W., Ageorges, N., Lidman, C., Moorwood, A.F.M., Spyromilio, J., Hubin, N., and Menten, K.M. (2002). A star in a 15.2-year orbit around the supermassive black hole at the centre of the Milky Way. *Nature*, **419**, 694–696.

Schödel, R. Ott, T., Genzel, R., Eckart, A., Mouawad, N., and Alexander, T. (2003). Stellar dynamics in the central arcsecond of our Galaxy. *Astrophys. J.*, **596**, 1015–1034.

Scoville, N.Z. (1972). Kinematics of gas near the Galactic Center. *Astrophys. J.*, **175**, L127–L132,

Serabyn, E., and Lacy, J.H. (1985). [Ne II] observations of the Galactic Center: Evidence for a massive black hole. *Astrophys. J.*, **293**, 445–448.

Shaver, P.A., Wall, J.V., Kellermann, K.I., Jackson, C.A., and Hawkins, M.R.S. (1996). Decrease in the space density of quasars at high redshift. *Nature*, **384**, 439–441.

Shklovsky, I.S. (1964). Nature of jets in radio galaxies. *Sov. Astron.*, **7**, 748–754.

Silk, J., and Rees, M. (1998). Quasars and galaxy formation. *Astron. Astrophys.*, **331**, 1–4.

Soltan, A. (1982). Masses of quasars. *Mon. Not. RAS*, **200**, 115–122.

Spitzer, L., and Saslaw, W.C. (1966). On the evolution of galactic nuclei. *Astrophys. J.*, **143**, 400–419.

Stark, A.A., Martin, C.L., Walsh, W.M., Xiao, K., and Lane, A.P. (2004). Gas density, stability, and starbursts near the inner Lindblad resonance of the Milky Way. *Astrophys. J.*, **614**, L41–L44.

Su, M., and Finkbeiner, D.P. (2012). Evidence for gamma-ray jets in the Milky Way. *Astrophys. J.*, **753:61**, 1–13.

Su, M., Slayter, T.R., and Finkbeiner, D.P. (2010). Giant gamma-ray bubbles from Fermi-LAT: Active Galactic nucleus activity or bipolar Galactic wind. *Astrophys. J.*, **724**, 1044–1082.

Sunyaev, R.A., Markevitch, M., and Pavlinsky, M. (1993). The Center of the Galaxy in the recent past: A view from GRANAT. *Astrophys. J.*, **407**, 606–610.

Tremaine, S. (1995). An eccentric-disk model for the nucleus of M31. *Astron. J.*, **110**, 628–633.

van der Kruit, P.C. (1970). Evidence for a possible expulsion of gas from the Galactic Nucleus. *Astron. Astrophys.*, **4**, 462–481.

van der Kruit, P.C., Oort, J.H., and Mathewson, D.S. (1972). The radio emission of NGC 4258 and the possible origin of spiral structure. *Astron. Astrophys.*, **21**, 169–184.

van Woerden, H., Rougoor, G.W., and Oort, J.H. (1957). Expansion d'une structure spirale dans le noyau du Systme Galactique, et position de la radiosource Sagittarius A. *Compt. Rend. l'Acad. Sci.*, **244**, 1691–1695.

Veilleux, S., Cecil, G., and Bland-Hawthorn, J. (2005). Galactic winds. *Ann. Rev. Astron. Astrophys.*, **43**, 769–826.

Viollier, R.D., Trautmann, D., and Tupper, G.B. (1993). Supermassive neutrino stars and galactic nuclei. *Phys. Lett. B.* **306**, 79–85.

Volonteri, M., and Bellovary, J. (2012). Black holes in the early Universe. *Rep. Prog. Phys.*, **75**, 124901.

Wardle, M., and Yusef-Zadeh, F. (1992). Origin of the hot gas and radio blobs at the Galactic Center. *Nature*, **357**, 308–310.

Wilson, A.S., and Ulvestad, J.S. (1982). Radio structures of Seyfert galaxies. IV – Jets in NGC 1068 and NGC 4151. *Astrophys. J.*, **263**, 576–594.

Wollman, E.R., Geballe, T.R., Lacy, J.H., Townes, C.H., and Rank, D.M. (1976). Spectral and spatial resolution of the 12.8 micron Ne II emission from the Galactic Center. *Astrophys. J.*, **205**, L5–L9.

Woltjer, L. (1959). Emission nuclei in galaxies. *Astrophys. J.*, **130**, 38–44.

Woltjer, L. (1964). A source of energy in radio galaxies. *Nature*, **201**, 803–804.

Zeldovich, Ya.B. (1964). The fate of a star and the evolution of gravitational energy upon accretion. *Sov. Phys. Doklady*, **9**, 195.

Index